Catalysis: Current and Future Developments

Volume 1

Fundamentals and Prospects of Catalysis

Edited by

Goutam Kumar Patra

Department of Chemistry,
Guru Ghasidas Vshwavidyalaya (A Central University),
Chhattisgarh 495009,
India

Santosh Singh Thakur

Department of Chemistry,
Guru Ghasidas Vshwavidyalaya (A Central University),
Chhattisgarh 495009,
India

Catalysis: Current and Future Developments

Volume # 1

Fundamentals and Prospects of Catalysis

Editors: Goutam Kumar Patra & Santosh Singh Thakur

ISSN (Online): 2737-4416

ISSN (Print): 2737-4408

ISBN (Online): 978-981-14-5851-4

ISBN (Print): 978-981-14-5849-1

ISBN (Paperback): 978-981-14-5850-7

Published by Bentham Science Publishers Pte. Ltd. Singapore. All Rights Reserved.

need for a court order if at any point you breach any terms of this License Agreement. In no event will any delay or failure by Bentham Science Publishers in enforcing your compliance with this License Agreement constitute a waiver of any of its rights.

3. You acknowledge that you have read this License Agreement, and agree to be bound by its terms and conditions. To the extent that any other terms and conditions presented on any website of Bentham Science Publishers conflict with, or are inconsistent with, the terms and conditions set out in this License Agreement, you acknowledge that the terms and conditions set out in this License Agreement shall prevail.

Bentham Science Publishers Pte. Ltd.
80 Robinson Road #02-00
Singapore 068898
Singapore
Email: subscriptions@benthamscience.net

BENTHAM SCIENCE

CONTENTS

PREFACE

Catalysis is one of the basic fundamental as well as thrust research areas of chemical sciences which fascinates a wide range of academicians, researchers, chemical technologists and industries throughout the world. The field of catalysis is interdisciplinary by its nature, includes organic synthesis, coordination and organometallic chemistry, kinetics and mechanism, stereochemical concepts and material science all at its very heart. Though the catalysis is an established area of research but one in which novel unexplored development continues exponentially. In the past few decades 2001, 2005 and 2010 nine Nobel laureates enriched on the domain of catalysis generating practical use of asymmetric catalysis to stereoselectivity of asymmetric hydrogenation and oxidation, development of metathesis method in organic synthesis and catalytic carbon-carbon coupling reactions respectively. Further, chemical and enzymatic catalysis is very popular and higly dynamic research area which received 15 times recognition from the Nobel Foundation during 1901 to 2012.

The downstream chemicals and valuable commercial products obtained via the diverse catalysts and catalytic processes have been gaining unprecedented prominence which could be evidenced by increasingly large numbers of publications, patents, monographs, symposia and conferences. Innumerable organic, inorganic and biochemical reactions are facilitated by catalysis and thus it plays a crucial role in chemical sciences. The global catalyst market shows about 90% of all commercially produced bulk and fine chemicals, biochemical, synthetic and medicinal products involve homogeneous or heterogeneous catalysis at some stage in the chemical or biochemical process of their manufacture. In recent years, there is tremendous growth in various types of subfields in catalysis e.g., nanocatalysis, asymmetric or chiral catalysis, industrial catalysis, organocatalysis, photocatalysis, electrochemical catalysis, enzyme and biocatalysis, tandem catalysis, autocatalysis, induced catalysis, and environmental catalysis. The quantitative requirement of the catalyst in a chemical process from the stoichiometric amount to catalytic amount and potential to change equilibrium excites the researcher to dream about perpetual motion machine, a contradiction to the law of thermodynamics. Stereospecific catalysts not only accelerate the reaction rates but also control the product absolute configuration, therefore a deep insight into kinetics and mechanism of catalysis is vital to get desired products conversion rates, turnover number and frequency of the catalytic reactions processes. Therefore, catalysis development and understanding is crucial not only to the academics or basic microscopic level but industrial, technology or macroscopic level as well.

This book involves the most distinctive characteristics of almost all-inclusive of the catalysis and highlights many important topics and subfields/subdisciplines. The novel design, synthesis, development, reducing energy consumption and side products, atom economy and green chemistry approach of catalysis definitely would play a vital role in future perspective of catalysis.

We hope this book will serve as an excellent reference for graduate students, researchers at all levels in both academic and industrial laboratories.

Goutam Kumar Patra
Department of Chemistry
Guru Ghasidas Vshwavidyalaya (A Central University)
Chhattisgarh 495009
India

&

Santosh Singh Thakur
Department of Chemistry
Guru Ghasidas Vshwavidyalaya (A Central University)
Chhattisgarh 495009
India

List of Contributors

Arti Shrivastava	Department of Chemistry, Guru Ghasidas Vishwavidyalaya, Bilaspur (Chhattsgarh) 495009, India
Bhaskar Sharma	Department of Chemistry, Guru Ghasidas Vishwavidyalaya, Bilaspur (Chhattsgarh) 495009, India
Deepak Patel	Department of Chemistry, Guru Ghasidas Vishwavidyalaya (A Central University), Koni, Bilaspur (Chhattisgarh) 495009, India
Geetika Patel	Department of Chemistry, Guru Ghasidas Vishwavidyalaya (A Central University), 495009 (Chhattisgarh), India
Goutam Kumar Patra	Department of Chemistry, Guru Ghasidas Vishwavidyalaya (A Central University), Koni, Bilaspur (Chhattisgarh) 495009, India
Hema Tandon	Department of Chemistry, Guru Ghasidas Vishwavidyalaya, Bilaspur (Chhattisgarh) 495009, India
Kalluri V.S. Ranganath	Department of Chemistry, Institute of Science, Banaras Hindu University, Varanasi, India
Kalpataru Das	Department of Chemistry, School of Chemical Sciences and Technology, Dr. Harisingh Gour University (A Central University), Sagar - 470003 (M.P.), India
Kavita Jain	Department of Chemistry, School of Chemical Sciences and Technology, Dr. Harisingh Gour University (A Central University), Sagar - 470003 (M.P.), India
Kiran Thakur	Department of Chemistry, Government Pataleshwar College Masturi, Bilaspur (Chhattisgarh) 495551, India
Medha Kiran Patel	Department of Chemistry, Guru Ghasidas Vishwavidyalaya (A Central University), 495009 (Chhattisgarh), India
Megha Balha	Department of Chemistry, Indian Institute of Technology Guwahati, Assam 781039, India
Melad Shaikh	Indian Institute of Science Education and Research, Tirupati, India
Nidhi Nirmalkar	Department of Chemistry, Guru Ghasidas Vishwavidyalaya (A Central University), Koni, Bilaspur (Chhattisgarh) 495009, India
Pratibha Mandal	Department of Chemistry, Guru Ghasidas Vshwavidyalaya (A Central University), Koni, Bilaspur (Chhattisgarh) 495009, India
Pathik Maji	Department of Chemistry, Guru Ghasidas Vishwavidyalaya, Bilaspur (Chhattisgarh) 495009, India
Rama Jaiswal	Department of Chemistry, Institute of Science, Banaras Hindu University, Varanasi, India
Santosh Singh Thakur	Department of Chemistry, Guru Ghasidas Vishwavidyalaya (A Central University), Koni, Bilaspur (Chhattisgarh) 495009, India
Subhas Chandra Pan	Department of Chemistry, Indian Institute of Technology Guwahati, Assam 781039, India
Subhash Banerjee	Department of Chemistry, Guru Ghasidas Vishwavidyalaya (A Central University), 495009 (Chhattisgarh), India

CHAPTER 1

Organocatalytic Asymmetric Synthesis of Spiroacetals and Bridged Acetals

Megha Balha and **Subhas Chandra Pan**[*]

Department of Chemistry, Indian Institute of Technology Guwahati, Assam 781039, India

Abstract: Chiral spiro compounds have been found to have great importance in organic synthesis because of their presence in a variety of natural alkaloids and pharmaceuticals. Lately, spiro compounds have gained interest because of their engrossing conformational characteristics and their structural connection with biological systems. Enantioselective synthesis of conformationally constrained spiro and bridged acetals can also be performed by organometallic catalysts but we will focus on organocatalytic routes in this chapter. Organocatalytic methodologies are found to be powerful approaches for the synthesis of conformationally rigid spiro and bridged acetal compounds because of their stability, functional group tolerance and easy stereoprediction. In this chapter, the organocatalytic asymmetric approaches for the synthesis of bridged [2.2.1], [2.2.2], [3.1.1], [3.2.1], [3.3.1] bicyclic acetals as well as spiroacetals are discussed in details with examples. The synthesis contains different reactions such as Michael addition reaction, Mannich reaction, cycloaddition reaction, aldol reaction, tandem Friedel Craft/hemiketalization reaction, Knoevenagel, Diels Alder reaction, cyclisation reaction and various other reactions.

Keywords: Asymmetric, Bridged acetal, Chiral Compounds, Enantioselectivity, Organocatalysts, Spiro Compounds.

INTRODUCTION

"The universe is asymmetric and I am persuaded that life, as it is known to us, is a direct result of the asymmetry of the universe or of its indirect consequences. The universe is asymmetric."- Louis Pasteur. The term asymmetric means lack of equality or equivalence between parts or aspects of something. Lack of equality or equivalence can be found anywhere like nature, chiral molecule or human body. Human bodies are asymmetrical, though they look symmetrical from outside, but most of our vital organs are positioned asymmetrically. Different organoleptic properties of enantiomers are shown by lemon and orange, respectively, containing the left- and right-handed version (enantiomers) of the same molecule,

---

[*] **Corresponding author Subhas Chandra Pan:** Department of Chemistry, Indian Institute of Technology Guwahati, Assam 781039, India; Tel: +91-361-258-3304; Fax: +91-361-258-2349; E-mail: span@iitg.ac.in

Goutam Kumar Patra & Santosh Singh Thakur (Eds.)

Limonene. Asymmetric synthesis is the synthesis of chiral compounds and it focuses on the production of one stereoisomer over another stereoisomer. Spiroacetals contain acetals connected to a spiro carbon atom and bridged acetals are acetals in which two ether groups are connected by a bridge.

Catalysts are used in the reaction to reduce the activation energy of the molecules. Organocatalysts are composed of carbon, hydrogen, sulphur and other non-metal elements. Organocatalysts have many advantages over traditional metal catalysts because:

1. They require mild reaction conditions.
2. They are stable and easy to design and synthesize.
3. They do not require anhydrous conditions, thus, reducing the cost of the synthesis.
4. They prevent the formation of metallic waste, thus are environment friendly.
5. They are compatible with several functional groups, thus no need to protect sensitive functional groups. Ultimately, reducing the number of reaction steps.

In the late 1890s, Von Bayer discovered spirocycles [1]. Spirocyclane is a term innovated by Von Bayer for naming bicyclic hydrocarbons having two rings with a carbon atom. In organic synthesis, the construction of spiro cyclic framework was quite challenging. Spiro compounds have attracted the attention of organic chemists especially in drug discovery in the last few decades because of their intrinsic complexity and rigidity. Many direct and efficient strategies have been developed for the synthesis of compounds having spiro scaffold. Spiro compounds are present in a plethora of natural products such as angiotensin antagonist irbesartan (1), analgesic morphine (2), β-vetivone (3), the antibiotic monensin (4), alkaloid citrinalin (5), opioid receptor agonist oxycodone (6), rosmadial (7), and (-)-acorenone B (8) (Fig. 1) [2].

Similarly, bridged *O,O*-acetals skeleton is present in a variety of natural products like procyanidin A1 (9) [3], cholinesterease inhibitor (11) [4], epicoccolide A (12) [5], and other bioactive compounds (Fig. 2) [6].

REVIEW OF INVESTIGATION RESULTS: ORGANOCATALYTIC ASYMMETRIC SYNTHESIS OF SPIROKETALS

In 2012, the Nagorny group reported a chiral phosphoric acid catalysed enantioselective as well as diastereoselective spiroketalization reaction of cyclic enol ethers bearing an alcohol in the alkyl chain. The reaction afforded spiroketals in good yields (up to 96%) with excellent stereoselectivities (Scheme 1) [7].

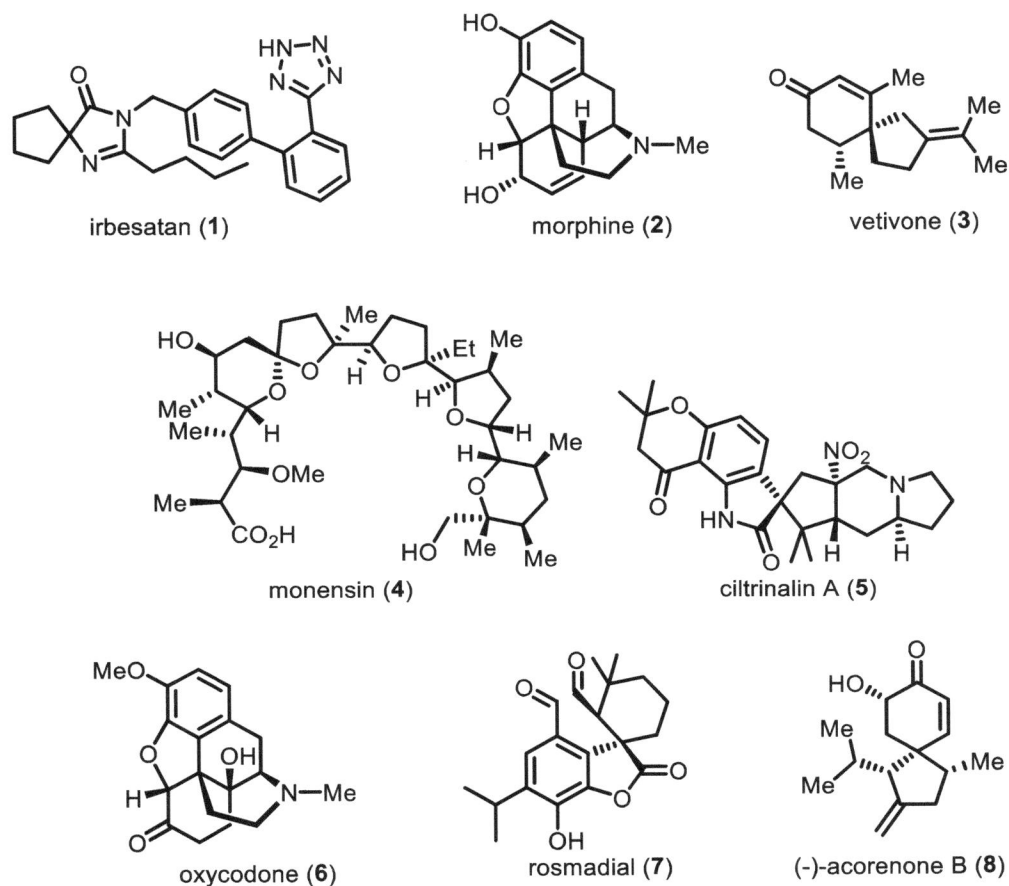

Fig. (1). Examples of natural products containing a spirocyclic ring.

Different chiral and achiral phosphoric acids were found to be effective in stimulating the reaction, but the best results were found with (S)-TRIP. Different solvents were also screened. Low levels of stereocontrol were observed in relatively polar solvents, whereas, in hydrocarbon solvents the reaction proceeded with higher enantioselectivities and shorter reaction time. Pentane was found to be the best solvent. The role of addition of 4Å MS was not clarified but it was considered important to isolate products with good to high levels of enantioselectivity (entry 3). Substrates with substituents in the aromatic rings and extension of the tether length did not affect the selectivity as well as the yield of the reaction. Moreover, substrates containing less rigid benzyl groups resulted in decreased conversion and enantioselectivity (entry 3).

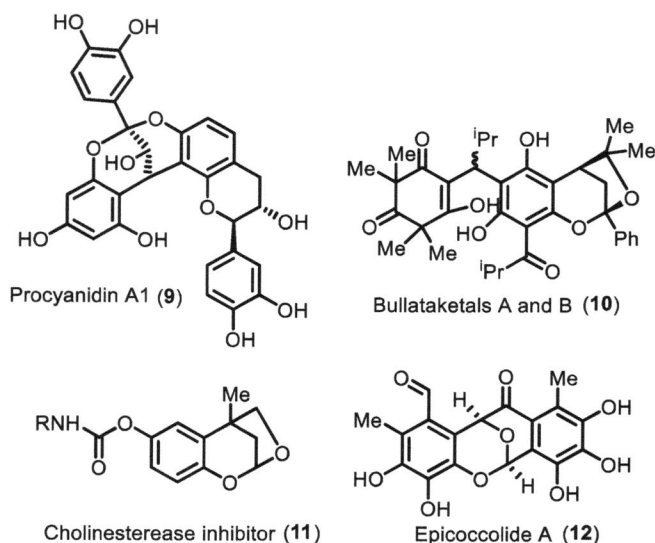

Procyanidin A1 (**9**)

Bullataketals A and B (**10**)

Cholinesterease inhibitor (**11**)

Epicoccolide A (**12**)

Fig. (2). Examples of natural products containing benzofused acetal scaffold.

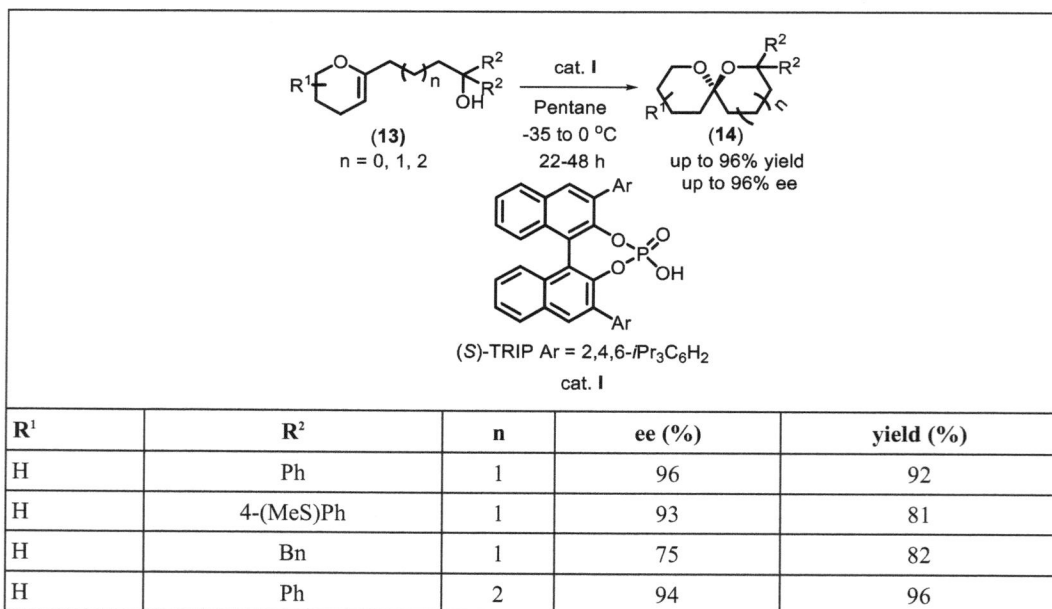

R¹	R²	n	ee (%)	yield (%)
H	Ph	1	96	92
H	4-(MeS)Ph	1	93	81
H	Bn	1	75	82
H	Ph	2	94	96

Scheme 1. Chiral phosphoric acid catalysed spiroketalization reaction.

Authors also successfully extended the scope to *D*-glucal derivatives for highly diastereoselective cyclization and excellent result was achieved (Scheme **2**) [7].

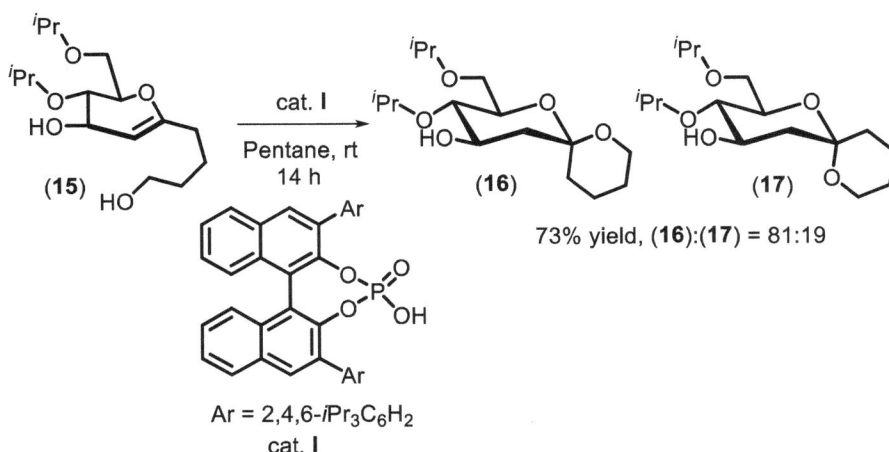

Scheme 2. Spiroketalization of cyclic enol ethers derived from sugars.

In the same year, List *et al.* reported an enantioselective spiroacetalization catalysed by conformationally locked Brønsted acids (Scheme **3**) [8]. The reaction afforded various spiroketals with good yields (up to 89%) and excellent stereoselectivities (up to >50:1 d.r. and up to 98:2 e.r.) using an imidodiphosphoric acid catalyst (cat. **II**). Remarkably, only 0.1 mol% of catalyst was required for the construction of the 5,5-spiroacetal (entry 2). Different enol ether ring sizes were also well tolerated to provide 7,6- and 7,5- spiroacetals by highly enantioselective reactions (entries 3 and 4). Kinetic resolution of racemate (**20**) delivered both bisacetal (**21**) and enolacetal (**22**) with excellent enantioselectivities (Scheme **4**) [8].

In 2015, Matsubara *et al.* reported an intramolecular hemiacetalization/oxy-Michael addition cascade catalysed by bifunctional amino-thiourea catalyst. This strategy resulted in the formation of spiroketal compounds bearing an alkyl group at the 2-position which are prevalent in insect hormones. The spiroketals were isolated in moderate to excellent yields (40-99%) with moderate to excellent stereoselectivities. This method was applied to electron-rich and electron-poor enones as well as aliphatic enones and products were isolated with good to excellent yields and enantioselectivities (Scheme **5**) [9]. Chalcogran (**27**), a pheromone of the six-spined spruce bark beetle *Pityogenes chalcographus*, was also synthesised using the obtained product (**26**) (Scheme **6**) [9].

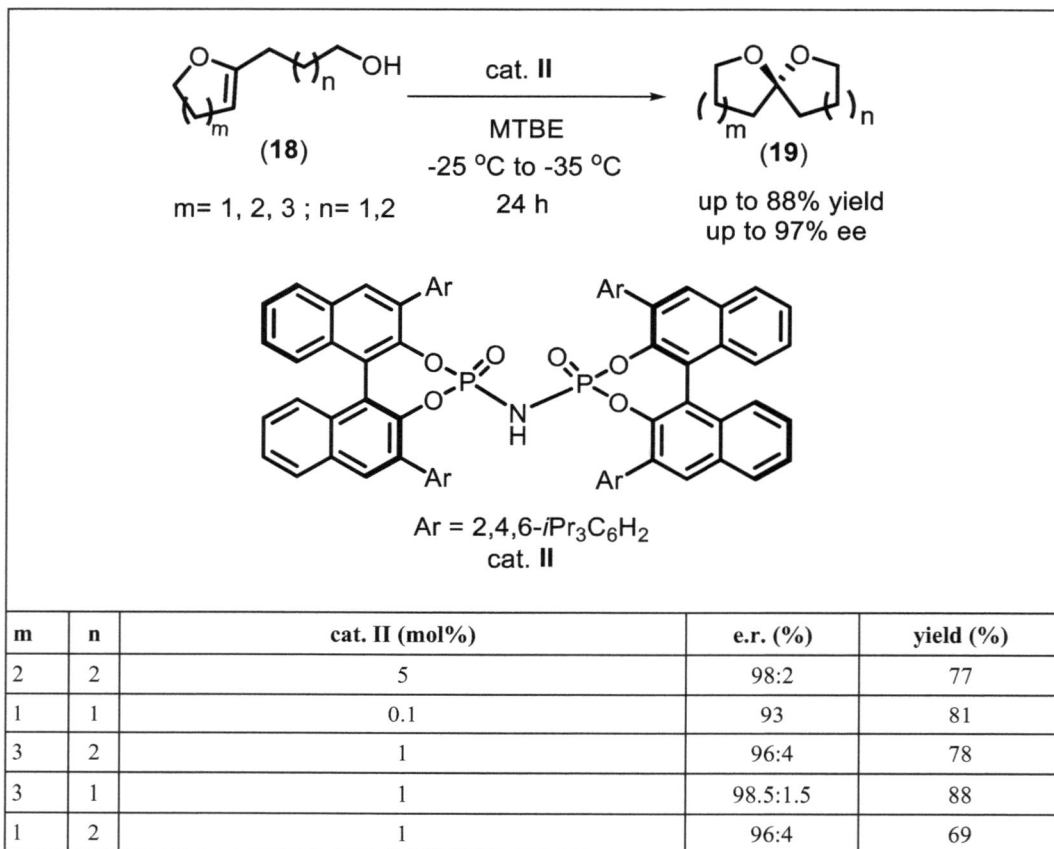

Scheme 3. Catalytic asymmetric spiroacetalization.

m	n	cat. II (mol%)	e.r. (%)	yield (%)
2	2	5	98:2	77
1	1	0.1	93	81
3	2	1	96:4	78
3	1	1	98.5:1.5	88
1	2	1	96:4	69

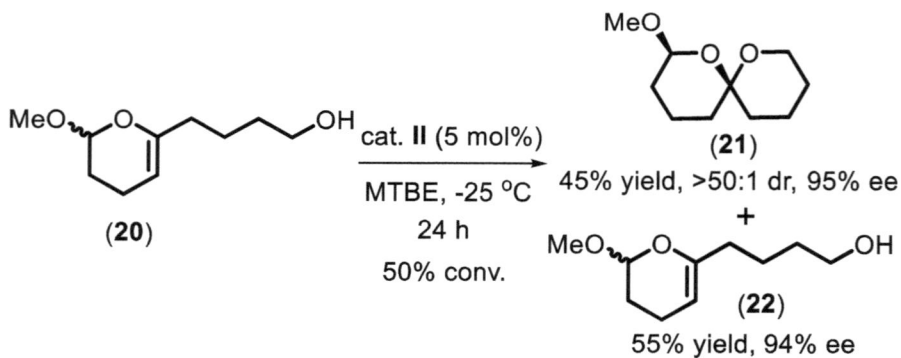

Scheme 4. Kinetic resolution of racemate (**20**).

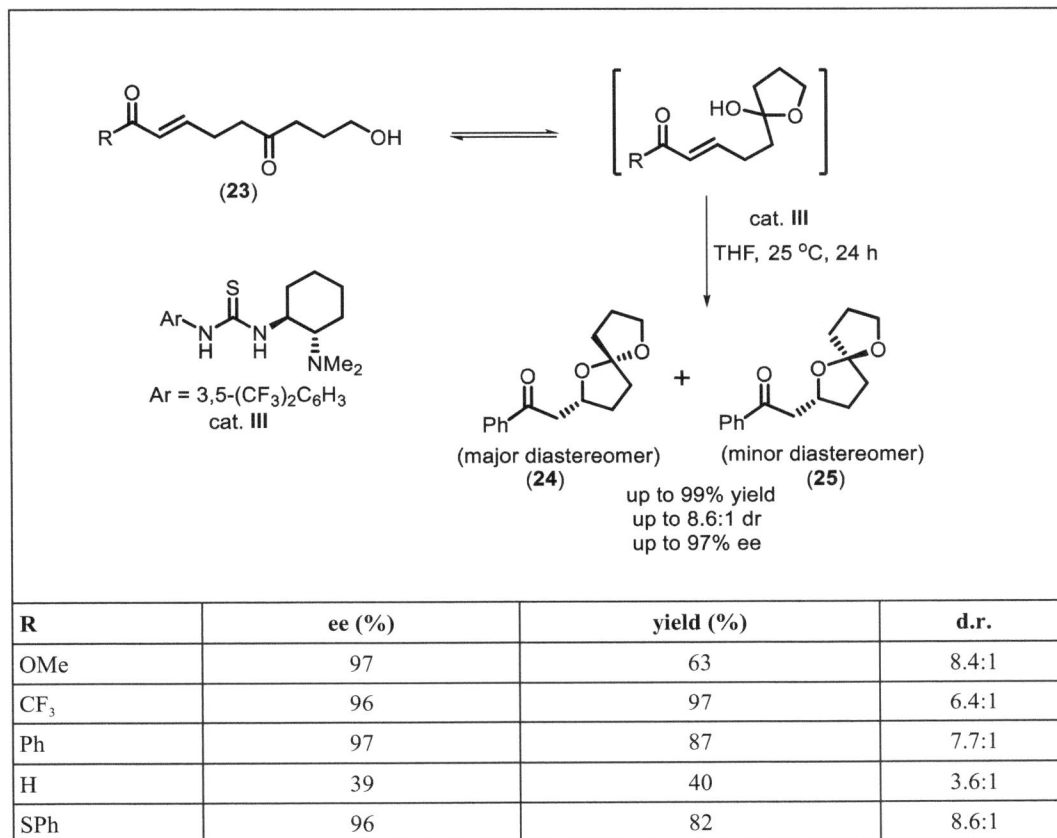

Scheme 5. Spiroketalization catalysed by amino-thiourea catalyst.

R	ee (%)	yield (%)	d.r.
OMe	97	63	8.4:1
CF₃	96	97	6.4:1
Ph	97	87	7.7:1
H	39	40	3.6:1
SPh	96	82	8.6:1

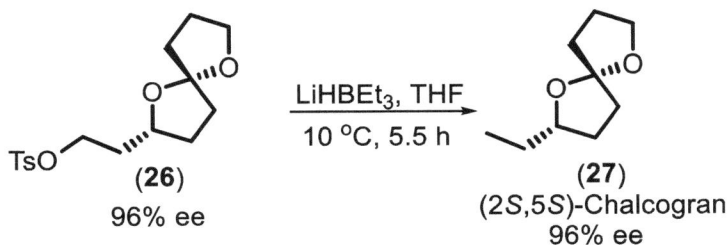

Scheme 6. Synthesis of (2S,5S)-Chalcogran.

In 2016, Xue, Jiang and Li *et al.* reported organohalogenite-mediated asymmetric intramolecular aromatic spiroketalization (Scheme **7**) [10]. Earlier reports focused on the synthesis of aliphatic spiroketals. However, in this report the authors emphasized the enantioselective construction of aromatic spiroketals, though it was quite challenging because of the lower nucleophilicity of aromatic hydroxyl groups. The reaction afforded bisbenzannulated spiroketals in moderate to good yields (52-92%) with excellent enantioselectivities. 1,3-dibromo-5,5-dimethyl-hydantoin (DBDMH) acts as a halogen source. Substrates with aryl rings bearing electron-withdrawing and electron-donating substituents were well tolerated in this method.

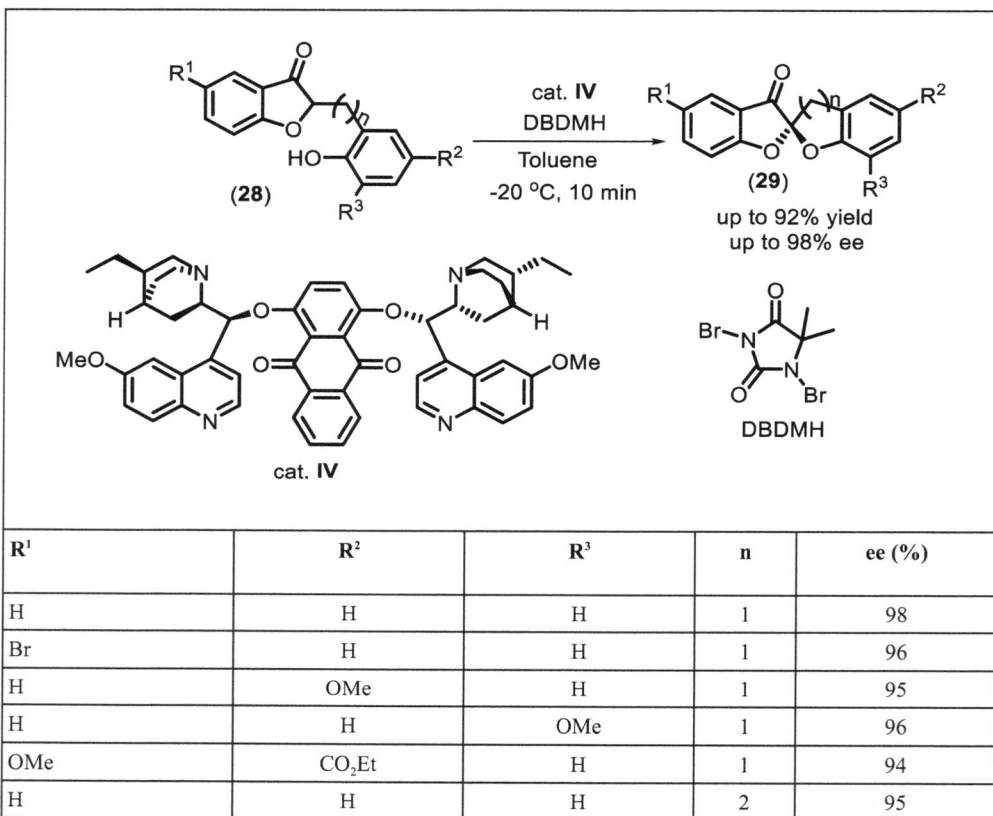

Scheme 7. Synthesis of Bisbenzannulated spiroketal cores.

R¹	R²	R³	n	ee (%)	yield (%)
H	H	H	1	98	70
Br	H	H	1	96	85
H	OMe	H	1	95	52
H	H	OMe	1	96	57
OMe	CO₂Et	H	1	94	92
H	H	H	2	95	74

(R)-2,2-Diphenyl-1,7-dioxaspiro[5.5]undecane (14)

General Procedure [7]

In a round bottom flask, starting material (**13**) (0.1 mmol), 4 Å molecular sieves (100 mg) and cat.**I** (0.005 mmol) were added. The mixture was cooled to −78 °C; stirred for 5 min and then 5 mL of pentane was added. The mixture was stirred for another 5 min before it was warmed up to −35 °C. The reaction mixture was stirred for a selected time and triethylamine was added to quench the reaction.

Purification was done by column chromatography (which was presaturated with triethylamine) to provide spiroketal (**14**); yield: 81-96%; ee: 74-96%.

(S)-1,6-Dioxaspiro[4.4]nonane (19)

General Procedure [8]

In a closed vial with a septum, solvent (7 ml) and molecular sieves were added and the mixture was cooled to an appropriate temperature. A solution of substrate (**18**) (0.25 mmol) in solvent (2 ml) was then added, and the mixture was stirred for 5-10 min allowing it to reach the reaction temperature. To the reaction mixture a solution of catalyst **II** in solvent (1 ml) was added dropwise. After designated time at the designated temperature, the reaction was quenched with Et$_3$N (50 µl). Purification was performed by column chromatography to afford product (**19**); yield: 62-89%; er: 98.5:1.5.

1-Phenyl-2-((2R,5S)-1,6-dioxaspiro[4.4]nonan-2-yl)ethan-1-One (24)

General Procedure [9]

To a 5 mL round bottom flask, substrate (**23**) (0.1 mmol), THF (0.2 mL), and cat. **III** (2.1 mg, 0.005 mmol) were sequentially added. The mixture was stirred in an oil bath at 25 °C for 24 h. The reaction mixture was diluted with hexane/EtOAc (v/v = 1/1), passed through a short silica gel pad to remove cat. **III**; concentrated in vacuo. Column chromatography (hexane/EtOAc (v/v = 1/1)) afforded the corresponding 2-alkylspiroketals (**24**) and (**25**); yield: 40-99%; ee: 39-97%; dr: 3.6:1-8.4:1.

(*S*)-3H,3'H-2,2'-Spirobi[benzofuran]-3-one (29)

General Procedure [10]

Compound (**28**) (0.2 mmol) and catalyst **IV** (0.04 mmol) in toluene (4.0 mL) were treated with DMDBH (0.12 mmol) at an appropriate temperature. The reaction mixture was allowed to stir for 10 min. The reaction mixture was filtered to remove the catalyst and the filtrate was concentrated in *vacuo* at room temperature. Column chromatography (Petroleum ether/EtOAc 16:1, v/v) gave the product (**29**); yield: 52-92%; ee: 92-98%.

REVIEW OF INVESTIGATION RESULTS: ORGANOCATALYTIC ASYMMETRIC SYNTHESIS OF BRIDGED ACETALS

In 2013, Franzén *et al.* reported the catalytic asymmetric synthesis of optically active *O,O*-acetals (Scheme **8**) [11]. The reaction was carried out with hydroxyl enals (**30**) and acetyl acetone (**31**) in the presence of a pyrrolidine catalyst (cat. **V**). Treatment of the crude mixture with 1.2 equivalents of AcCl/BF$_3$·OEt$_2$ resulted in the formation of acetal **32** whereas, with TFA, the formation of enol was found. Thus, this reaction has a built in chemoselective switch simply by changing the acid that allows the selective formation of either acetal or enol ether. The reaction afforded *O,O*-acetal derivatives in good yields and excellent stereoselectivities. Unsymmetrical diketones provided the desired product as a single regioisomer in moderate yield whereas 1:1 mixture of diastereoisomers was observed with prochiral cyclohexadione.

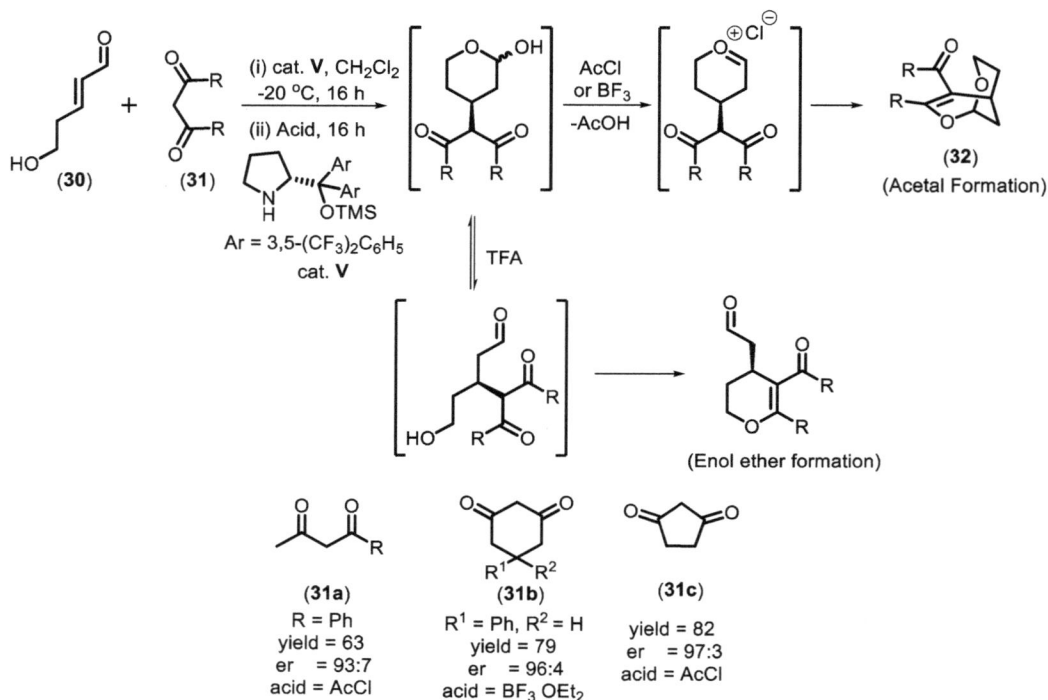

Scheme 8. Enantioselective synthesis of *O,O*-acetals.

In 2016, Karl Anker Jørgensen *et al.* reported an enantioselective organocatalytic reaction between γ-keto-enals and 1-naphthols for the synthesis of methanobenzodioxepine (**35**) and tetrahydrofurobenzofuran (**36**) scaffolds catalysed by secondary amine catalysts (Scheme **9**) [12]. The reaction proceeded via two reaction pathways: the first path led to the formation of chiral 5,6-bridged methanobenzodioxepine scaffolds (**35**) containing three stereocenters, whereas, the other pathway provided 5,5-fused tetrahydrofurobenzofuran scaffolds (**36**) bearing two stereocenters. The methanobenzodioxepines as well as tetrahydrofurobenzofurans were formed in moderate to good yields with excellent enantioselectivities. The formation of tetrahydrofurobenzofuran and methanobenzodioxepine scaffolds was highly dependent on the nature of substituents in the γ-keto-enal (Scheme **9**). When R^1 = -Ph, -H then methanobenzodioxepine scaffolds will form whereas, if R^1 = -Et, -iPr, Bn, n-pentane then tetrahydrofurobenzofuran scaffolds will form.

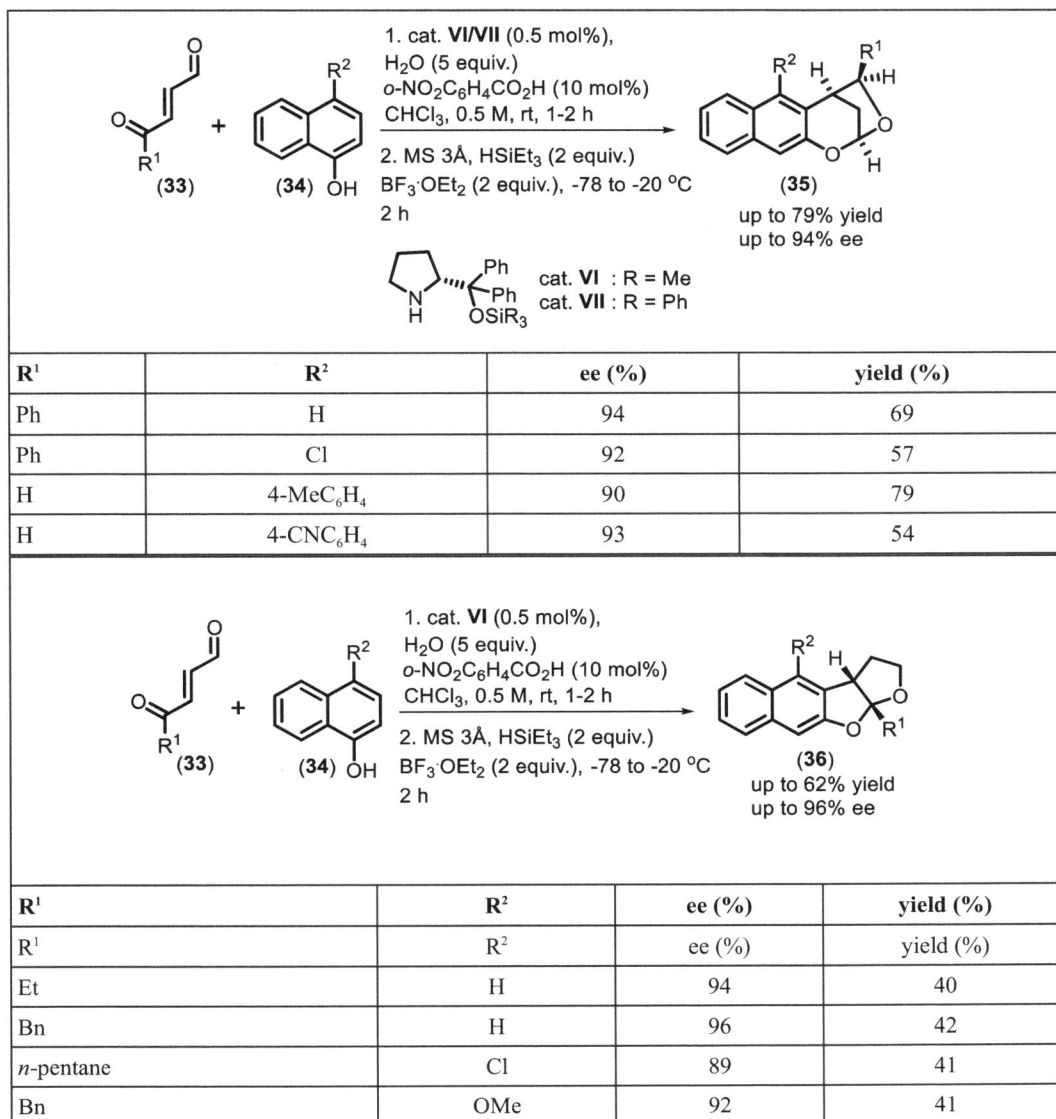

R^1	R^2	ee (%)	yield (%)
Ph	H	94	69
Ph	Cl	92	57
H	4-MeC$_6$H$_4$	90	79
H	4-CNC$_6$H$_4$	93	54

R^1	R^2	ee (%)	yield (%)
R^1	R^2	ee (%)	yield (%)
Et	H	94	40
Bn	H	96	42
n-pentane	Cl	89	41
Bn	OMe	92	41

Scheme 9. Enantioselective synthesis of benzofused acetals.

In 2018, Pan *et al.* reported an organocatalytic asymmetric reaction between *ortho*-hydroxy-cinnamaldehyde (**37**) and *N*-benzyl dioxindole (**38**) catalysed by a secondary amine catalyst (scheme **10**) [13]. This method involved amine catalyzed conjugated addition followed by diastereoselective acetalization with TFA which led to the formation of spirooxindole products (**39**); obtained in good to high yields with high diastereo- and enantioselectivities. The electronic effects of the substituents did not influence the outcome of the reaction much.

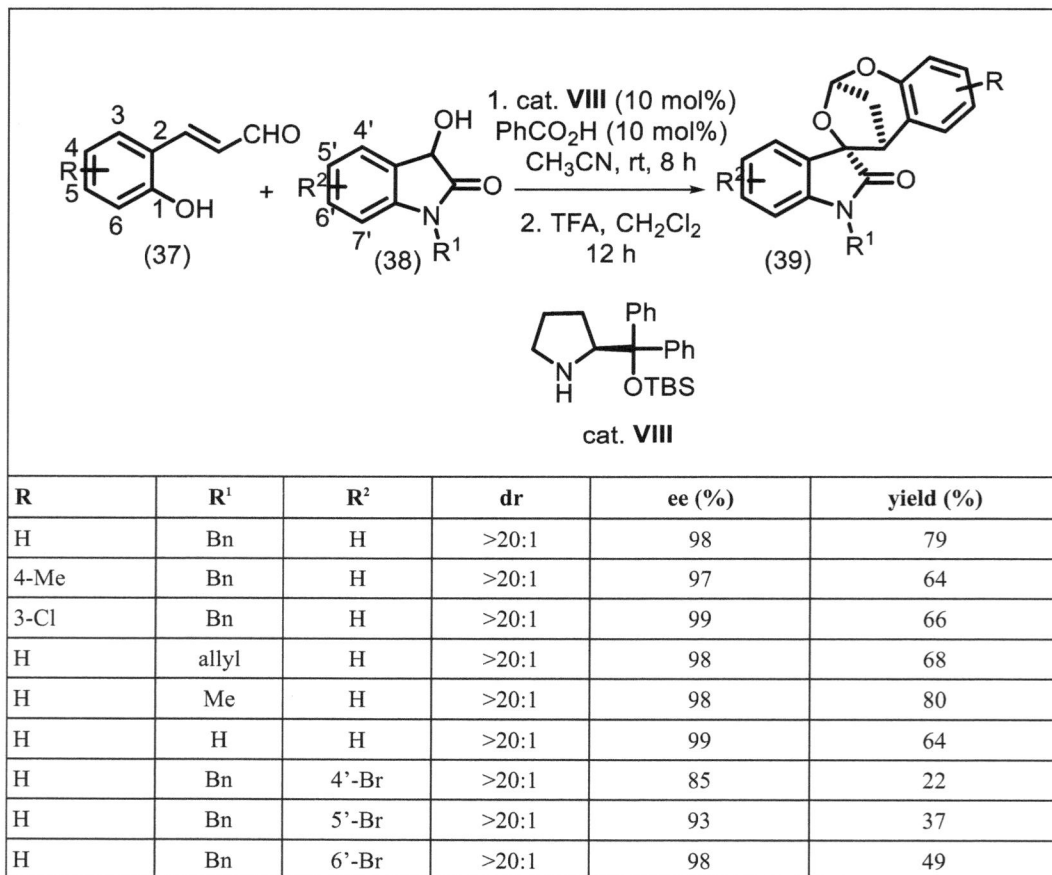

R	R¹	R²	dr	ee (%)	yield (%)
H	Bn	H	>20:1	98	79
4-Me	Bn	H	>20:1	97	64
3-Cl	Bn	H	>20:1	99	66
H	allyl	H	>20:1	98	68
H	Me	H	>20:1	98	80
H	H	H	>20:1	99	64
H	Bn	4'-Br	>20:1	85	22
H	Bn	5'-Br	>20:1	93	37
H	Bn	6'-Br	>20:1	98	49

Scheme 10. Catalytic asymmetric synthesis of bridged acetals.

1-((1S,5S)-3-Methyl-2,8-dioxabicyclo[3.3.1]non-3-en-4-yl)ethan-1-one (32)

General Procedure [11]

In an oven-dried round-bottom, (*E*)-5-Hydroxypent-2-enal (**30**) (0.4 mmol) was added to a solution of diketone (**31**) (0.4 mmol) and cat. **V** (0.04 mmol) in CH_2Cl_2 (0.4 mL) at -20 °C. After full conversion of the starting materials, the acid was added to the reaction mixture at the given temperature. After stirring for 1 h, an aqueous saturated solution of Na_2CO_3 was added to quench the reaction, and the water phase was extracted with CH_2Cl_2. The combined organic phase was dried over Na_2SO_4, filtered, and concentrated under reduced pressure. Column chromatography (pentane/Et_2O) afforded the desired product (**32**); yield: 63-82%; er: 93:7-98:2.

Synthesis of Benzofused Acetals (35) and (36)

General Procedure [12]

A glass vial (4 mL) equipped with a magnetic stirring bar was charged with $CHCl_3$ (300 μL), hydroxyarene (**34**) (0.25 mmol, 1.0 eq.), o-$NO_2C_6H_4CO_2H$ (0.025 mmol, 0.1 eq.), water (22.5μL,1.25 mmol, 5.0 eq.) and a 0.063 M solution of cat. **VI/VII** in $CHCl_3$ (0.0125 mmol, 0.005 eq.). Afterwards, the γ-keto-enal (**33**) (0.375 mmol, 1.5 eq.) was added in one portion and stirred at rt. Completion of the reaction was monitored by ^1H NMR. 3Å Molecular sieves (300 mg) and the NMR sample containing $CHCl_3$(500 μL) were added to the crude mixture and stirred for 2 h. Afterwards, $HSiEt_3$ (0.5 mmol, 2.0 eq.) was added, and the reaction mixture was cooled to -78°C. A solution of $BF_3 \cdot OEt_2$ in $CHCl_3$ (61.7 μL in 1 mL, 0.5 mmol, 2.0 eq.) was then added. The reaction mixture was allowed to warm up to -20°C and stirred at this temperature for 2 h. The solution was quenched with $NaHCO_3$ (aq) (5mL) and extracted with CH_2Cl_2 (3×5mL). The combined organic phases were dried with Na_2SO_4, filtered, concentrated in vacuo and the crude product was purified by FC. The purifications were initially performed over Iatrobeads; however, it was later discovered that the benzofused acetals were stable towards silica gel.

((2′R,3S,5′S)-1-Benzyl-5′H-spiro[indoline3,4′ [2, 5]methanobenzo[d] [1, 3] dioxepin]-2-one) (39)

General Procedure [13]

First step: In an oven-dried round-bottom flask, compound (**37**) (0.12 mmol), compound (**38**) (0.1 mmol), 10 mol % of cat. **VIII**, 10 mol % of $PhCO_2H$, and 0.5 mL of toluene were added. The reaction mixture was stirred at rt for 20 h. Completion of the reaction was checked by TLC. After completion of the reaction, the solution was concentrated and the residue was directly treated with TFA. Second step: DCM was added to the reaction mixture of first step and allowed to stir at rt. Then, 1.5 equiv. of TFA was added and the reaction mixture was allowed to stir overnight. Progress of the reaction was monitored by TLC. After the completion of reaction, solvent was concentrated and reaction mixture was directly purified by column chromatography on silica gel eluting with hexane/ethyl acetate (10%) to afford desired product (**39**); yield: 22-80%; 85-99% ee; >20:1 dr.

CONCLUSION

Conformationally rigid spiro and bridged acetal compounds are pharmaceutically important and possess structurally unique framework, which have drawn enormous attention from different research groups as well as from industries. Different novel methodologies have been developed in the last few years for the synthesis of these constrained structures, which we have discussed. In this chapter, we have shown, Michael addition reaction, Mannich reaction, cycloaddition reaction, aldol reaction, tandem Friedel Craft/hemiketalization reaction, Knoevenagel, Diels Alder reaction, and cyclisation reaction methods that have been used for this purpose. Thus, this chapter shows the potential of spiro and bridged acetal compounds for the development of new drugs and biologically active compounds.

CONSENT FOR PUBLICATION

Not applicable.

CONFLICT OF INTEREST

The author declares that there is no conflict of interest in this chapter.

ACKNOWLEDGEMENTS

Declared none.

REFERENCES

[1]   Baeyer, A.V. Systematik und Nomenclatur bicyclischer Kohlenwasserstoffe. *Ber. Dtsch. Chem. Ges.,* **1900**, *33*, 3771-3775.
        [http://dx.doi.org/10.1002/cber.190003303187]

[2]   Ding, A.; Meazza, M.; Guo, H.; Yang, J.W.; Rios, R. New development in the enantioselective synthesis of spiro compounds. *Chem. Soc. Rev.,* **2018**, *47*(15), 5946-5996.
        [http://dx.doi.org/10.1039/C6CS00825A] [PMID: 29953153]

[3]   Dumontet, V.; Van Hung, N.; Adeline, M.T.; Riche, C.; Chiaroni, A.; Sévenet, T.; Guéritte, F. Cytotoxic flavonoids and alpha-pyrones from *Cryptocarya obovata. J. Nat. Prod.,* **2004**, *67*(5), 858-862.
        [http://dx.doi.org/10.1021/np030510h] [PMID: 15165150]

[4]   Talontsi, F.M.; Dittrich, B.; Schüfler, A.; Sun, H.; Laatsch, H. Epicoccolides: Antimicrobial and Antifungal Polyketides from an Endophytic Fungus *Epicoccum* sp. Associated with *Theobroma cacao. Eur. J. Org. Chem.,* **2013**, 3174-3180.
        [http://dx.doi.org/10.1002/ejoc.201300146]

[5]   Kamal, M.A.; Qu, X.; Yu, Q-S.; Tweedie, D.; Holloway, H.W.; Li, Y.; Tan, Y.; Greig, N.H. Tetrahydrofurobenzofuran cymserine, a potent butyrylcholinesterase inhibitor and experimental

Alzheimer drug candidate, enzyme kinetic analysis. *J. Neural Transm. (Vienna),* **2008**, *115*(6), 889-898.
[http://dx.doi.org/10.1007/s00702-008-0022-y] [PMID: 18235987]

[6]   aHao, X-J.; Nie, J-L. Diterpenes from *Spiraea japonica. Phytochemistry,* **1998**, *48*, 1213-1215.
[http://dx.doi.org/10.1016/S0031-9422(97)00791-7] bShen, Z.; Chen, Z.; Li, L.; Lei, W.; Hao, X. Antiplatelet and antithrombotic effects of the diterpene spiramine Q from *Spiraea japonica* var. incisa. *Planta Med.,* **2000**, *66*(3), 287-289.
[http://dx.doi.org/10.1055/s-2000-8571] [PMID: 10821062] cLi, L.; Shen, Y-M.; Yang, X-S.; Zuo, G-Y.; Shen, Z.Q.; Chen, Z-H.; Hao, X-J. Antiplatelet aggregation activity of diterpene alkaloids from *Spiraea japonica. Eur. J. Pharmacol.,* **2002**, *449*(1-2), 23-28.
[http://dx.doi.org/10.1016/S0014-2999(02)01627-8] [PMID: 12163102]

[7]   Sun, Z.; Winschel, G.A.; Borovika, A.; Nagorny, P. Chiral phosphoric acid-catalyzed enantioselective and diastereoselective spiroketalizations. *J. Am. Chem. Soc.,* **2012**, *134*(19), 8074-8077.
[http://dx.doi.org/10.1021/ja302704m] [PMID: 22545651]

[8]   Čorić, I.; List, B. Asymmetric spiroacetalization catalysed by confined Brønsted acids. *Nature,* **2012**, *483*(7389), 315-319.
[http://dx.doi.org/10.1038/nature10932] [PMID: 22422266]

[9]   Yoneda, N.; Fukata, Y.; Asano, K.; Matsubara, S. Asymmetric Synthesis of Spiroketals with Aminothiourea Catalysts. *Angew. Chem. Int. Ed.,* **2015**, *54*, 15497-15500.
[http://dx.doi.org/10.1002/anie.201508405]

[10]   Xue, J.; Zhang, H.; Tian, T.; Yin, K.; Liu, D.; Jiang, X.; Li, Y.; Jin, X.; Yao, X. Organohalogenite-Catalyzed Spiroketalization: Enantioselective Synthesis of Bisbenzannulated Spiroketal Cores. *Adv. Synth. Catal.,* **2016**, *358*, 370-374.
[http://dx.doi.org/10.1002/adsc.201500390]

[11]   Polat, M.F.; Hettmanczyk, L.; Zhang, W.; Szabo, Z.; Franzén, J. One-Pot, Two-Step Protocol for the Catalytic Asymmetric Synthesis of Optically Active N,O- and O,O-Acetals. *ChemCatChem,* **2013**, *5*, 1334-1339.
[http://dx.doi.org/10.1002/cctc.201200860]

[12]   Paz, B.M.; Klier, L.; Naesborg, L.; Lauridsen, V.H.; Jensen, F.; Jørgensen, K.A. Enantioselective organocatalytic cascade approach to different classes of benzofused acetals. *Chem. Eur. J.,* **2016**, *22*(47), 16810-16818.
[http://dx.doi.org/10.1002/chem.201602992] [PMID: 27593532]

[13]   Balha, M.; Pan, S.C. Organocatalytic asymmetric synthesis of bridged acetals with spirooxindole skeleton. *J. Org. Chem.,* **2018**, *83*(23), 14703-14712.
[http://dx.doi.org/10.1021/acs.joc.8b02156] [PMID: 30372074]

CHAPTER 2

Design and Development of Bimetallic Enantioselective Salen Co Catalysts for the Hydrolytic Kinetic Resolution of Terminal Epoxides

Santosh Singh Thakur[1],*, **Deepak Patel[1]**, **Nidhi Nirmalkar[1]**, **Kiran Thakur[2]** and **Goutam Kumar Patra[1]**

[1] *Department of Chemistry, Guru Ghasidas Vishwavidyalaya (A Central University), Koni, Bilaspur (Chhattisgarh) 495009-India*

[2] *Department of Chemistry, Government Pataleshwar College Masturi, Bilaspur (Chhattisgarh) 495551-India*

Abstract: The hydrolytic kinetic resolution of terminal epoxides catalyzed by the monometallic chiral salen Co complex follows the cooperative bimetallic mechanism and second order kinetic dependency on the catalyst. In this mechanism, one metal works as an active Lewis acid center for preferential activation of one enantiomer from a racemic substrate and the second metal center stimulates the incoming nucleophile. Mechanistically, rational design and development of bi- and multimetallic chiral complex centers within the sterically, electronically, and co-ordinatively accessible framework of chiral salen ligand provides improved activity and enantioselectivity relative to their corresponding monometallic catalysts. This chapter provides a survey of bimetallic chiral salen Co complexes used in the hydrolytic kinetic resolution of terminal epoxides to procure valuable chiral intermediates, useful for academic interest and in industrial applications.

Keywords: Asymmetric catalysis, Chiral salen Co complex, Kinetic resolutions, Terminal epoxides.

BACKGROUND AND MOTIVATION

Asymmetric synthesis of chiral compounds, using enantioselective catalysts, plays a crucial role in chemical and pharmaceutical sciences [1]. The industrial need and great academic interest of enantiomerically pure (chiral) compounds in drug

---
* **Corresponding author Santosh Singh Thakur:** Department of Chemistry, Guru Ghasidas Vishwavidyalaya (A Central University), Koni, Bilaspur (Chhattisgarh) 495009-India; Tel: +91-9981209738 and +91-7999047064; E-mails: santosh.chirality@gmail.com, ss.thakur71@ggu.ac.in

Goutam Kumar Patra & Santosh Singh Thakur (Eds.)

synthesis have motivated, in the past decades, a tremendous growth in catalytic stereoselective synthesis [2 - 9]. A number of homogeneous chiral catalysts have already gained wide acceptance in terms of efficiency and selectivity, some of which are even used on an industrial scale, and the chemists involved in the pioneering breakthroughs were recently awarded the Nobel Prize [9]. Asymmetric or chirality (molecular handedness) is considered to be a major element in nearly all naturally occurring or manmade molecules that imparts a key role in science, engineering, and technology. There must be a fine-tuning of chirality in a wide range of chemical, biological and physical functions that are generated through precise stereochemical communications and molecular recognition. Life itself depends on chiral recognition because living systems, particularly enzymes, interact with enantiomers in decisively different manners [8].

Enantiomerically pure compounds can be obtained by one of three strategies [3]:

a. By resolution, either spontaneous (the way Pasteur resolved a tartaric acid salt) or with the aid of an enantiopure reagent. In a classical resolution, two diastereoisomers are formed, and their properties are sufficiently different so they can be separated by a conventional method, such as fractional crystallization or chromatography. The desired enantiomer is then obtained from one of the purified diastereoisomers, while the other is recycled, used for another purpose, or discarded. Classical methods are frequently applied on a large scale [2]. The method based on kinetic resolution is also frequently used.
b. By the use of chiral substrates ("chirons"), mainly of natural origin, which undergo highly stereoselective transformations leading to desired enantiomeric targets [2c].
c. By conversion of prochiral precursor into a chiral products (*i.e.* asymmetric synthesis). Biochemically, asymmetric synthesis can be performed by enzymes. Chemically, asymmetric synthesis can be performed using chiral auxiliaries, reagent, or catalysts [4].

These three methods have been actively developed for the past few decades. The asymmetric synthesis process should have high (greater than 90%) regio-, diastereo- and enantioselectivities. Furthermore, other important factors in this process are the expense and accessibility of the reagents, the conditions of the reaction (solvent, temperature, and pressure), and the ease of workup and purification. Keeping in mind all these practical considerations, less selective methods are sometimes favored over more selective ones, particularly for large-scale asymmetric synthesis.

Epoxides are classified as a member of ether family, carrying partially positive charged carbons and a Lewis-basic oxygen atom in a three-member ring system which may be terminal, di, tri, and tetra-substituted found in many interesting natural products (Scheme **1**) [10]. The characteristic polarity caused by the ring strain makes epoxides more reactive to countless organic reactions with a large number of reagents. Epoxides readily undergo nucleophilic ring-opening reactions, with hydrolysis and alcoholysis and allow straightforward elaboration to useful new functionality and often with the generation of new carbon-carbon bonds. Indeed, reactions of epoxides with nucleophiles, Lewis acids, radicals, reducing agents, oxidizing agents, acids, and bases have all been utilized in synthesis [11]. Therefore, the preparation of enantiomerically pure epoxides particularly, terminal epoxides, is very important because these compounds are attractive chiral building blocks for asymmetric synthesis.

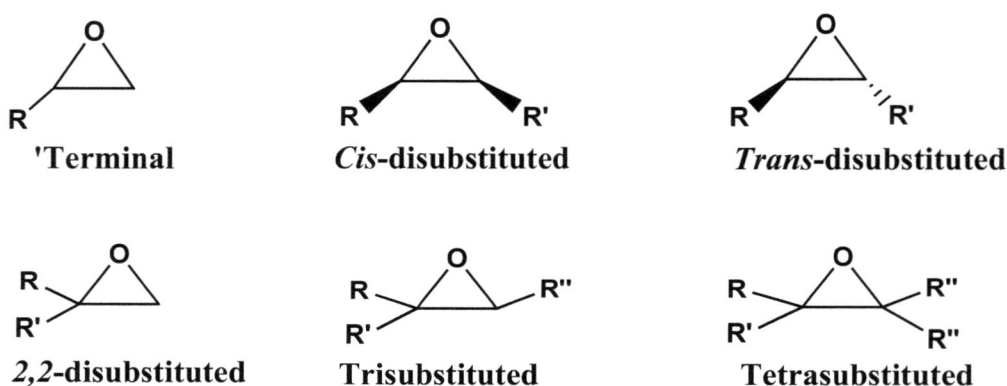

'Terminal   *Cis*-disubstituted   *Trans*-disubstituted

2,2-disubstituted   **Trisubstituted**   **Tetrasubstituted**

Scheme 1.

Among the above epoxides, enantioenriched1-oxiranes (terminal epoxides) are versatile compound (C-3 chiral building block) and the most valuable class of epoxides for organic synthesis. Naturally occurring epoxides are typically complex compounds and available only in limited amounts, Nature's chiral pool has not demonstrated to be a useful direct source of optically active epoxides for use in organic synthesis. Alternatively, enantioenriched epoxides can be prepared indirectly from the chiral pool *via* multistep procedures [12]. However, the range of epoxides available by this approach is also quite limited and inefficient. As a result, the preparation of optically pure epoxides has long been practiced as the most significant target for asymmetric synthesis. Especially, the identification of catalytic asymmetric olefin oxidation methods has been an area of potential research for several decades, and the advances made in this field have significantly increased the number of enantiomerically enriched epoxides available for use in organic synthesis [13].

The Katsuki-Sharpless epoxidation reaction, for the preparation of enantioenriched epoxides, among available techniques, has been found to very powerful and efficient methods of any asymmetric catalytic reaction explored thus far, providing general access to highly enantioenriched epoxyalcohols (Scheme **2a**) [14].

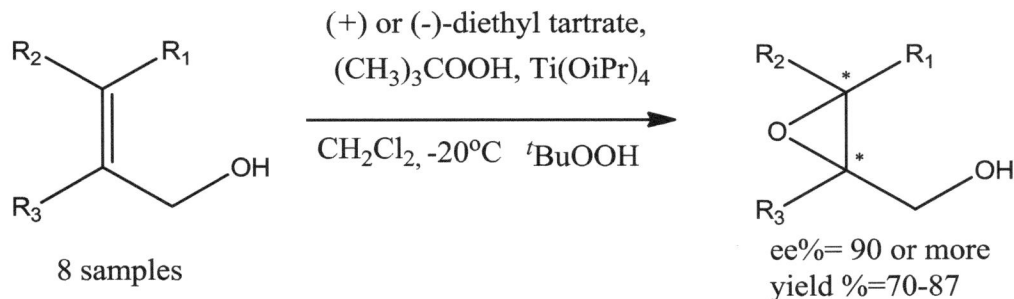

Scheme 2a.

The epoxidation of unfunctionalized olefins by chiral(salen)Mn^{III} complexes (Scheme **2b**) has made available the practical synthesis of certain classes of enantiomerically enriched epoxides [15].

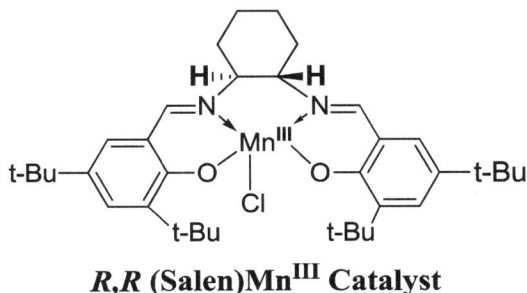

***R,R* (Salen)Mn^{III} Catalyst**

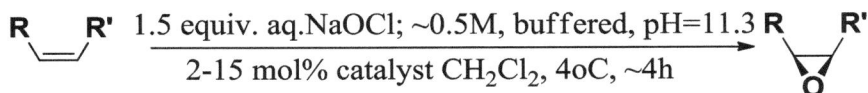

6 samples

R= Ar, Alkyl

R'= Alkyl (bulky group)

Scheme 2b.

Indirect routes to enantiopure epoxides involving asymmetric catalytic dihydroxylation or reduction reactions have also shown highly valuable in specific contexts [16]. Shi *et al.* reported the synthesis of optically active epoxides from various alkenes using a D-fructose-derived ketone organocatalyst with oxone as the primary oxidant and also reported the aziridination reactions [16f-g]. Shibasaki *et al.* reported the enantioselective synthesis of 2,2-disubstituted terminal epoxides *via* catalytic asymmetric Corey-Chaykovsky epoxidation of ketones catalyzed by heterobimetallic La-Li$_3$-BINOL complex. The resultant products 2,2-disubstituted terminal epoxides were obtained in high enantioselectivity (up to ee% 97) and yield (up to >99%) from a broad range of methyl ketones with 1–5 mol% catalyst loading with reaction time 12 to 60 h (Scheme **2c**) [16h]. The same research group also reported the addition of *tert*-butyl thiol to cyclic and acyclic epoxides in ee% =82-97 (yield%= 64-89%) *via* asymmetric ring opening reaction catalyzed by(*R*)-Ga-Li-bis(binaphthoxide) complex [16i].

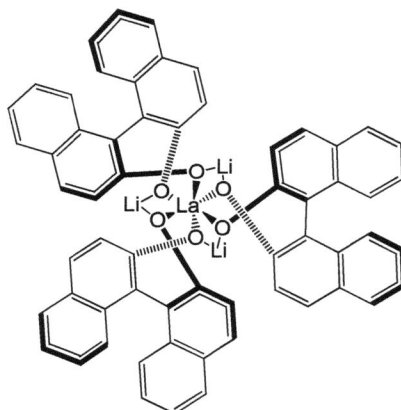

(*S*)-LaLi$_3$tris(binaphthoxide) (LLB).Catalyst

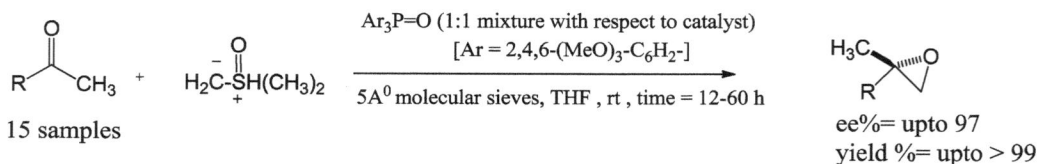

cat. (*S*)-LLB (1-5 mol%)
Ar$_3$P=O (1:1 mixture with respect to catalyst)
[Ar = 2,4,6-(MeO)$_3$-C$_6$H$_2$-]
5A^0 molecular sieves, THF , rt , time = 12-60 h

15 samples

ee%= upto 97
yield %= upto > 99

Scheme 2c.

Despite these considerable advances in the asymmetric catalytic synthesis of epoxides, there are very few reports on the direct preparation of highly enantioenriched1-oxiranes [17].

In this chapter, preparations of optically pure terminal epoxides have been discussed using bimetallic and multimetallic chiral (salen) Co complex as catalyst.

KINETIC RESOLUTION AND JACOBSEN HKR METHOD

Kinetic resolution method involves the principle of differential reaction rates (k_{rel}) of the chemical reaction of one of the enantiomers in a racemic mixture [18]. The selectivity factor of one enantiomer in a racemic mixture undergoes either the presence of a chiral catalyst (organic/organometallic) or chiral reagent (acid/base) or a biocatalyst (enzyme or microorganism) resulting in an enantioenriched sample of the less reactive enantiomer.

The enantiomeric excess (*ee*) of the unreacted starting material increases continually as more and more reactants converted to product, tending to 100% just before the end of the reaction. The kinetic resolution relies on the different chemical reactivity of the racemic mixture while the chiral resolution is based on the different physical properties of diastereomeric products (Scheme 3).

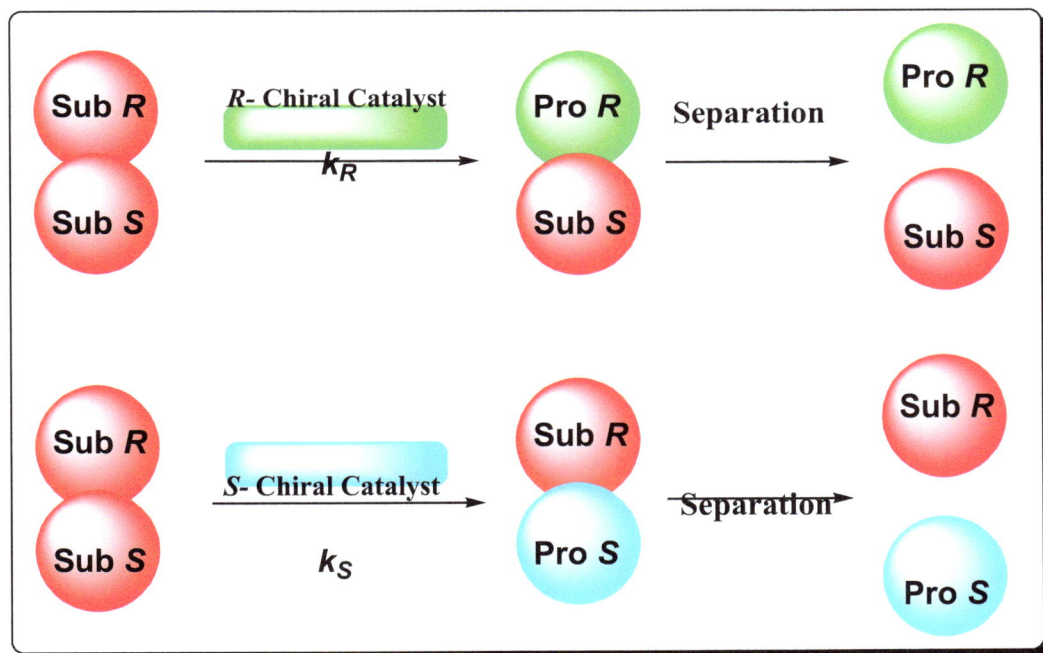

Scheme 3.

The basic condition to proceed for a kinetic resolution is when the rate constants for both the isomers are different *i.e.* $k_R \neq k_S$. In this method activation energies,

$\Delta G\ddagger$ or transition state energies of pairs of enantiomers are irreversibly different ($\Delta G_R^{\ddagger} \neq \Delta G_S^{\ddagger}$) even if the Gibbs free energy level of reactants or products are alike. In Scheme **4**, the value of $\Delta G_R^{\ddagger}$ is low for *R*-enantiomer, therefore, it reacts faster than the *S*-enantiomer. In catalytic asymmetric reactions, the magnitude of $\Delta\Delta G^{\ddagger}$ which stands for the difference in activation energies of high and low energy transition state of *S* and *R*-enantiomers respectively, depends upon the k_{rel} [19].

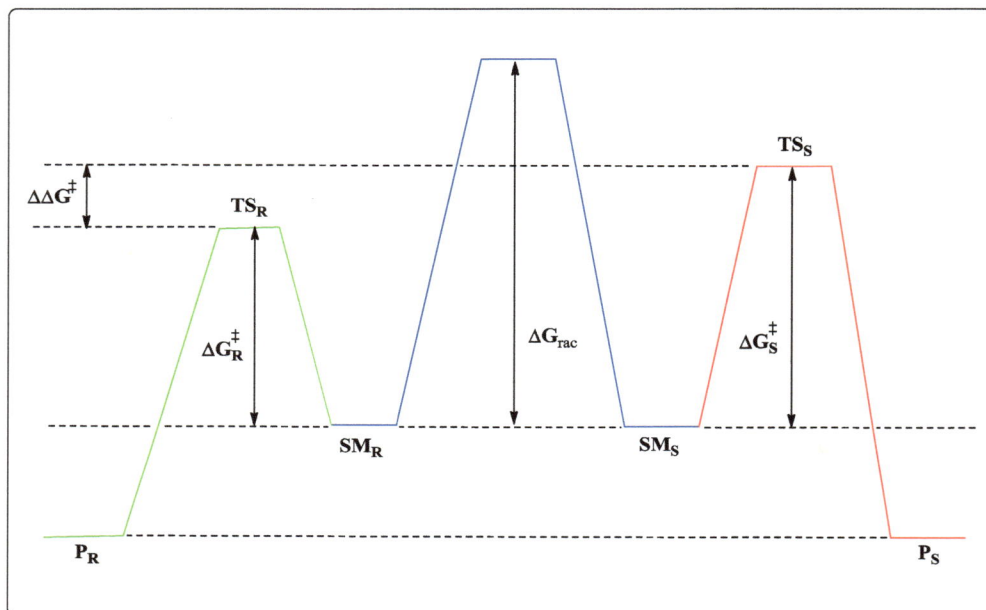

Scheme 4.

The selectivity(s) which is also referred to as the relative rates of reaction of a kinetic resolution, correlated with k_R and k_S respectively, by s=k_R/k_S, when $k_R>k_S$. This can be expressed mathematically in terms of the free energy difference as follows:

$$k_{rel} = \frac{k_{fast}}{k_{slow}} = e^{\Delta\Delta G^{\ddagger}/RT}$$

By assuming the first-order kinetics of substrate, the selectivity can also be expressed in terms of ee of the recovered starting material and conversion (*c*). If it is assumed that the *S* enantiomer of the starting material racemate reacts slower and will be recovered in excess, it is possible to express the concentrations (mole fractions) of the *S* and *R* enantiomers as:

$$[S] = \frac{(1 + ee)(1 - c)}{2}$$

$$[R] = \frac{(1 + ee)(1 - c)}{2}$$

Jacobsen and co-workers developed a methodology for the kinetic resolution of epoxides particularly terminal epoxides *via* nucleophilic ring-opening with the attack by an azide anion [20], carboxylic acid [21], water [22], phenols [23], *etc.* catalyzed by chiral salen Cr or Co complex. The chiral chiral (salen)Cr-N_3 catalyst displayed highly regioselective and enantioselective ring opening terminal epoxides *via* kinetc resolution path with $TMSN_3$ [20]. The ring opened products are used to prepare valuable optically active 1,2-amino-2-alkanols precursors. The k_{rel} for various terminal epoxides have been observed in the range from 44 to 230 which provides high level of enantio-differentiation by the catalyst for better kinetic resolution within reaction time from 18 to 50 hours in solvent free condition (Scheme **5a**). This catalyst is also powerful to catalyze the resolution of 2,2-disbustituted epoxides which reaction was not possible by salen) Co^{III}-(OAc) complex.

Jacobsen *et al.* firstly reported the kinetic resolution of racemic terminal epoxides with water catalyzed by chiral (salen)Co^{III}-OAc catalyst (Scheme **5b**) to get both terminal epoxides and their corresponding diols in the highly enantiopure form [22]. Therefore, asymmetric hydrolytic kinetic resolution (HKR)developed by the same research group has been proven to be one of the most powerful approaches to synthesize both products with high optical purity. The Hydrolytic Kinetic Resolution (HKR) of terminal epoxides now used to prepare the enantio-enriched terminal epoxides on laboratory and industrial scale [24]. A racemic terminal epoxide is combined with approximately half an equivalent of water in the presence of 0.2–2 mol% of a chiral Co(salen) catalyst proceeded at room temperature to afford a mixture of unreacted epoxide and 1,2-diol in almost equal amounts [24]. Either enantiomer of the epoxides can be obtained using different enantiomers of the (salen)Co^{III}-(OAc) complex.

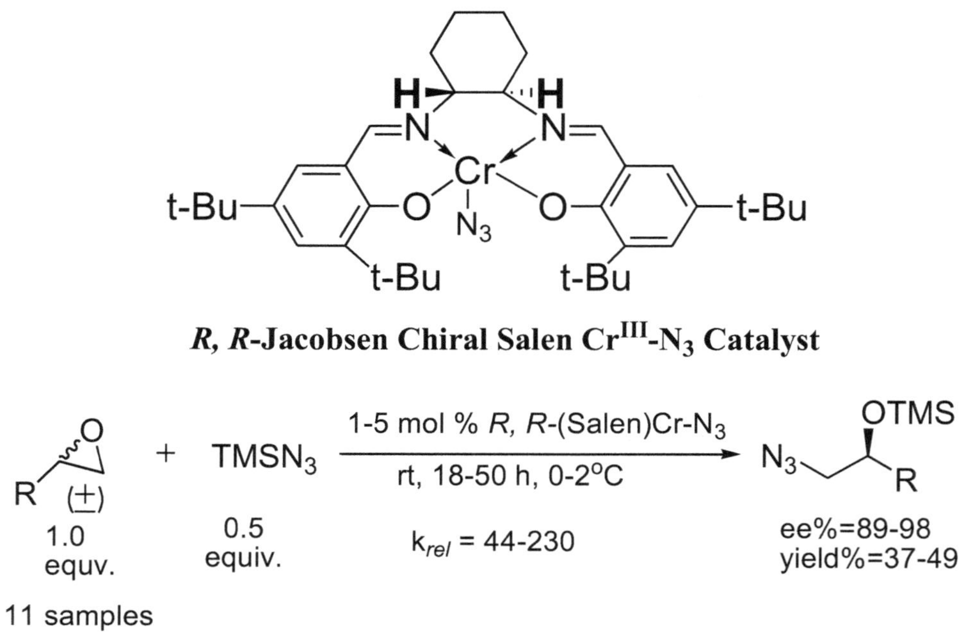

R, R-Jacobsen Chiral Salen Cr^{III}-N$_3$ Catalyst

Scheme 5a.

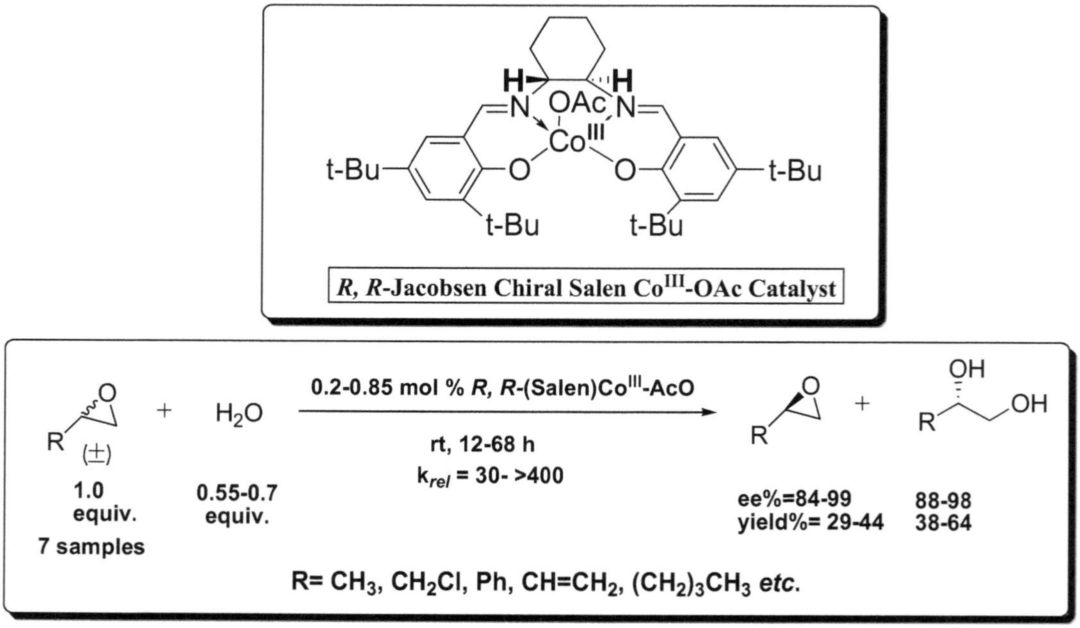

R, R-Jacobsen Chiral Salen Co^{III}-OAc Catalyst

R= CH$_3$, CH$_2$Cl, Ph, CH=CH$_2$, (CH$_2$)$_3$CH$_3$ *etc.*

Scheme 5b.

A broad spectrum of racemic terminal epoxides bearing a wide range of functional groups can afford the corresponding chiral epoxides with >99% ee through this process. They have also reported a mechanistic basis for high reactivity of (salen)Co−OTs in the hydrolytic kinetic resolution of terminal epoxides [25].

The comparision of catalytic performance chiral (salen)Co−OAc and (salen)Cr-N$_3$ complexes in the kinetic resolutions of terminal epoxides is shown in Table **1** (though the reaction parameters were not identical) which reveals the superior activity and enantioselectivity of chiral (salen)Co−OAc even under low loading amount of catalyst.

Table 1. Comparision of the catalysts performances of chiral (salen)Cr-N$_3$ and chiral (salen)Co-OAc in asymmetric kinetic resolution of therminal epoxides.

No.	Catalyst/Amount (mol%)	R	Nucleophile/ Equivalent	React.condition/ Time (h)	ee (%)/ Isolated Yield	k_{rel}	Ref.
1	(salen)Cr-N$_3$/1.0	CH$_3$	(CH$_3$)$_3$SiN$_3$/0.5	0-2°C/18-50 h	97/49	230	[20]
	(salen)Co-OAc/0.2		H$_2$O/0.55	room temp/12 h	98/50	>400	[22a]
2	(salen)Cr-N$_3$/2.0	(CH$_2$)$_3$ CH$_3$	(CH$_3$)$_3$SiN$_3$/0.5	0-2°C/18-50 h	97/44.5	160	[20]
	(salen)Co-OAc/0.42		H$_2$O/0.55	room temp/5 h	98/48	290	[22a]
3	(salen)Cr-N$_3$/2.0	CH$_2$Cl	(CH$_3$)$_3$SiN$_3$/0.5	0-2°C/18-50 h	95/47	100	[20]
	(salen)Co-OAc/2.0		H$_2$O/0.5	THF/4°C/24 h	96/50	218	[22b]

The hydrolytic kinetic resolution (HKR) catalyzed by (salen) Co(III) complex shows second-order dependence on Co catalyst and follows a cooperative bimetallic mechanism for the HKR analogous and to other asymmetric ring-opening (ARO) reactions with (salen) metal complexes (Eq 1.) [26].

It displayed that catalytic activity and selectivity depends upon the number of active chiral [Co] units and activation of the nucleophile. Based on this fact, several bimetallic and multimetallic (dendrimer and oligomeric) chiral salen Co catalysts have been designed and developed for the asymmetric ring-opening reactions of *meso* and terminal epoxides with various nucleophiles, including water *i.e.* hydrolytic kinetic resolution method [27].

$$\boxed{\text{Co}}\text{—Nu} \; + \; \triangleright\!\!\text{O}\text{—}\boxed{\text{Co}} \quad \longrightarrow \quad \text{Ring Opened Product} \qquad (1)$$

BI- AND MULTIMETALIC HYDROLYTIC KINETIC RESOLUTION

There has been tremendous develoment in the area of asymmetric hydrolytic kinetic resolution of terminal epoxides catalyzed by chiral (salen)Co since its first report by Jacobsen *et al.* in 1997 [22]. Some representative examples of mono-,b--, oligo- and, multimetallic chiral (salen)Co are summarized in Table **2**. The detailed description of each examples is beyond the scope of this chapter therefore, selected cases are described based on metal/active centre of the chiral salen catalyst in the HKR reactions.

Table **2**. Some representative Examples and development of the chiral (salen)Co/Cr-X catalyst in hydrolytic kinetic resolution (HKR)/kinetic resolution reaction of terminal epoxides.

No.	Metal Type/number of Active Centre	Mono-, bi-, Oligo-, Multimetallic	Catalyst Amount (mol %)	Counter Ion X=	Substrate $\overset{O}{\underset{R}{\triangle}}$ R=	Recovered Substrate ee (%)/ Yield (%)[c]	Ring Opened Product ee (%)/ Yield(%)	Reference (s)
1	Cr/one	Monometallic	1.0	N_3	CH_3	N.A.	97/49	[20]
2	Co/one	Monometallic	0.2	OAc	CH_3	.98/ 44	98/50	[22]
3	Co/four	Oligomeric	0.0003	OTf	CH_3	99/40	N.A.	[27, 31]
4	Co/eight	Dendrimer	0.027	I	*Cyclo*-C_6H_{11}	98/50	N.A.	[28a]
5	Co/four	Oligomeric	0.0004	NBS[a]	CH_3	>99/45	97/51	[30]
6	Co/four	Multimetallic	0.01	OTf	C_4H_9	>99/NA	N.A.	[32]
7	Co/two	Bimetallic	0.4	Cl	CH_3	>99/NA	N.A	[33a]
8	Co/two	Bimetallic	0.5	Cl	C_2H_5	>99/45	86/42	[34]
9	Co/two	Bimetallic	0.8	Cl	C_6H_5	98/36	N.A.	[35]
10	Co/two	Bimetallic	0.2	Cl	H_3C-O-CH_2	99.7/41	86.5/40	[36]
11	Co/two	Bimetallic	0.1	Cl	CH_2Cl	99.8/45	87.1/43	[37b]
12	Co/two	Bimetallic	0.2	OTf	CH_3	99.4/42	89.4/41	[43]
13	Co/two	Bimetallic	0.02	OAc	CH_2Cl	99/51	N.A.	[47]
14	Co/two	Bimetallic	0.04	OAc	CH_2Cl	99/52	N.A.	[47]
15	Co/two to ten	Multimetallic	0.01	OAc	CH_2Cl	99/50.3	N.A.	[49]

(Table 2) cont.....

No.	Metal Type/number of Active Centre	Mono-, bi-, Oligo-, Multimetallic	Catalyst Amount (mol %)	Counter Ion X=	Substrate R R=	Recovered Substrate ee (%)/ Yield (%)[c]	Ring Opened Product ee (%)/ Yield(%)	Reference (s)
16	Co/five to ten	Multimetallic	0.01	OAc	C_3H_5-O-CH_2	>99/50.4	N.A.	[51]
17	Co/one	Multimetallic	0.157 wt%	OAc	CH_3	98/50	98/NA	[52]
18	Co/one	Multimetallic	0.02	OAc	CH_2OBn	>99/53.5	86.9/NA	[53]
19	Co/one	Multimetallic	0.5	OAc	CH_2Cl	>99/52	91.9/NA	[54]
20	Co/one	Multimetallic	0.1	OTf	CH_2Cl	>99/39	N.A.	[55]
21	Cr/one	Multimetallic	0.5	OTs	n-C_6H_{13}	>99/46	>95/50.	[59]
22	Co/One	Multimetallic	0.5	OAc	n-Bu	92/50	N.A.	[60]
23	Co/two	Bimetallic	0.3	OTs	n-Bu	>99.9/50.4	91.3	[61]
24	Co/two	Bimetallic	0.03	OTs	$(CH_2)_3CH_2$	99/41	N.A.	[62]
25	Co/four	Multimetallic	0.7	OAc	CH_2O (1-naphthyl)	99/57	N.A.	[63]
26	Co/not defined	Multimetallic	2.0	OAc	CH_2Br	86/80[d]	N.A.	[66a]
27	Co/three	Multimetallic	0.01	OAc	CH_2Cl	>99/52	N.A.	[67]
28	Co/two	Bimetallic	0.1	OAc	CH_2Cl	>99/52	N.A.	[69]
29	Co and Mn /two	Multimetallic	1.0	OAc and Cl	CH_2Br	92/97[d]	N.A.	[72]
30	Co/two	Bimetallic	0.3	OAc	CH_2Cl	98/42	86/50	[73a]
31	Co/two	Bimetallic	0.1	OAc	CH_2Cl	>99/35	75/60	[73b]
32	Co/one	Monometallic	0.025	PNBA[b]	$C_6H_5OCH_2$	.99/ 49	99/48	[74]
33	Co/one	Monometallic	0.2	not defined	CH_3	N.A.	98/>49	[77]

[a]NBS=N-bromosuccinimide
[b]PNBA= p-nitrobenzoic acid
[c] Conversion more than 50% either estimated by formula $ee_{epoxide}$ /ee_{diol} / [1+ ($ee_{epoxide}$ /ee_{diol})] on the basis of the ee value of the recovered epoxide and diol products or water/nucleophile is taken more than 0.5 equiv. with respect to substrate *i.e. rac* epoxide
[c] N.A. Data not available in the mentioned literature under identical reaction condition
[d]dynamic kinetic resolution

Jacobsen research group developed cooperative asymmetric catalysis with dendrimeric [Co(Salen)] complexes using highly pure polyamidoamine (PAMAM) and poly (propyleneimine) with four, eight and sixteen chiral Lewis acid Co centers [28a]. The dendritic catalyst (*R, R*)- 8-CoPAMAM (Scheme **6**) catalyst provided 98% ee of epoxide with a maximum 50% conversion in the

HKR reaction of (*rac*) vinyl cyclohexane epoxide taking as low as 0.027 mol % of the catalyst.

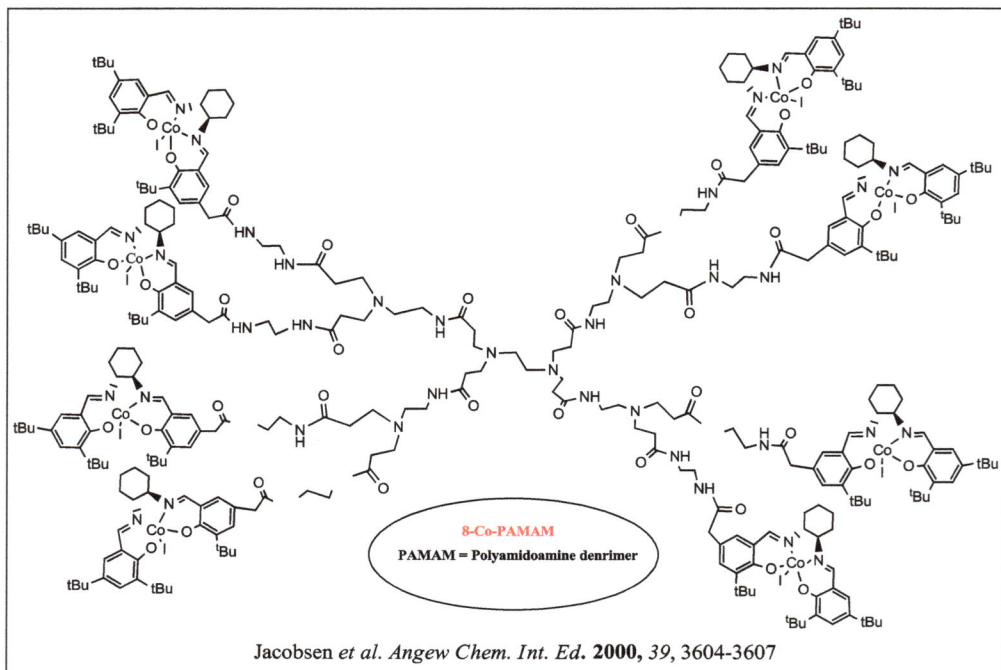

Jacobsen *et al. Angew Chem. Int. Ed.* **2000**, *39*, 3604-3607

Scheme 6.

The application of the oligomeric Co(salen) catalyst system (Scheme **7**) can further increase the reactivity to up to 50-fold per Co(salen) unit basis [29 - 31]. The HKR of propylene oxide was especially efficient: 1.5 mol of epoxide was resolved within 24 h using an extremely low amount of catalyst *i.e.* 3.6 mg (41 ppm by mass, 3 ppm on a molar basis) of catalyst to provide 35 g of recovered enantiopure propylene oxide [31].

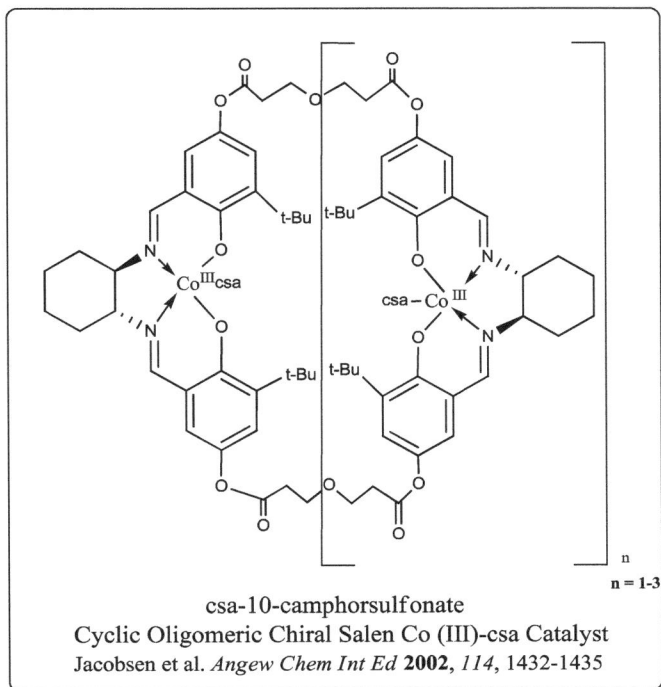

csa-10-camphorsulfonate
Cyclic Oligomeric Chiral Salen Co (III)-csa Catalyst
Jacobsen et al. *Angew Chem Int Ed* **2002**, *114*, 1432-1435

Scheme 7.

However, 2,2-disubstituted epoxides are un-reactive under Jacobsen's HKR conditions with this chiral catalyst and multistep and complicated synthesis method of oligomeric catalysts and solubility problems during HKR make it less attractive strategy. Further, they immobilized the chiral (salen)Co^{III} complexes on gold colloids [32]. The long chain thiolates functionalized gold colloids with end group of chiral [(salen)Co (III) were shown as an excellent catalyst in the hydrolytic kinetic resolution (HKR) of hexene-1-oxide and displayed extremely high selectivity (>99.9% ee) and reactivity (within 5 h) by 0.01mol% catalyst loading amount. To achieve the same result by homogeneous monomeric Jacobsen catalyst's (-OTf counter ion in place of -OAc Scheme **5**) higher loading amount of the catalyst (>0.1mol%) and more reaction time (within 24 h) was required. Further, the recycling of the immobilized catalyst can be done by simple filtration. The recycled catalyst can be re-oxidized and re-used (seven times) in the above HKR reaction with no loss of reactivity or enantioselectivity.

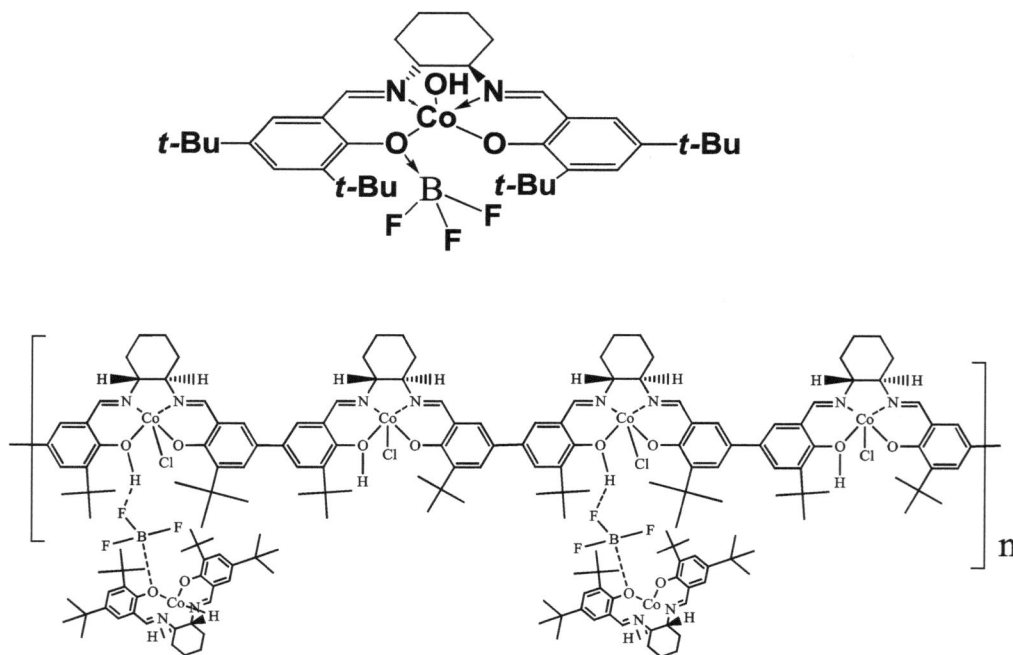

Scheme 8.

Kim *et al.* [33] reported the activation of inactive chiral (salen)Co^{II} to Co^{III} using Lewis acid $BF_3 \cdot 2H_2O$ and $BF_3 \cdot (C_2H_5)_2O$ in methylene chloride solvent and forming monometallic and polymetallic salen Co complexes (Scheme **8**) at open atmosphere, where aerobic oxidation took place by molecular oxygen. The amount of 0.4 mol% of the polymetallic chiral salen Co(III) catalyst showed high activity and enantioselectivity than the monomeric counterpart in the HKR reaction of various terminal epoxides.

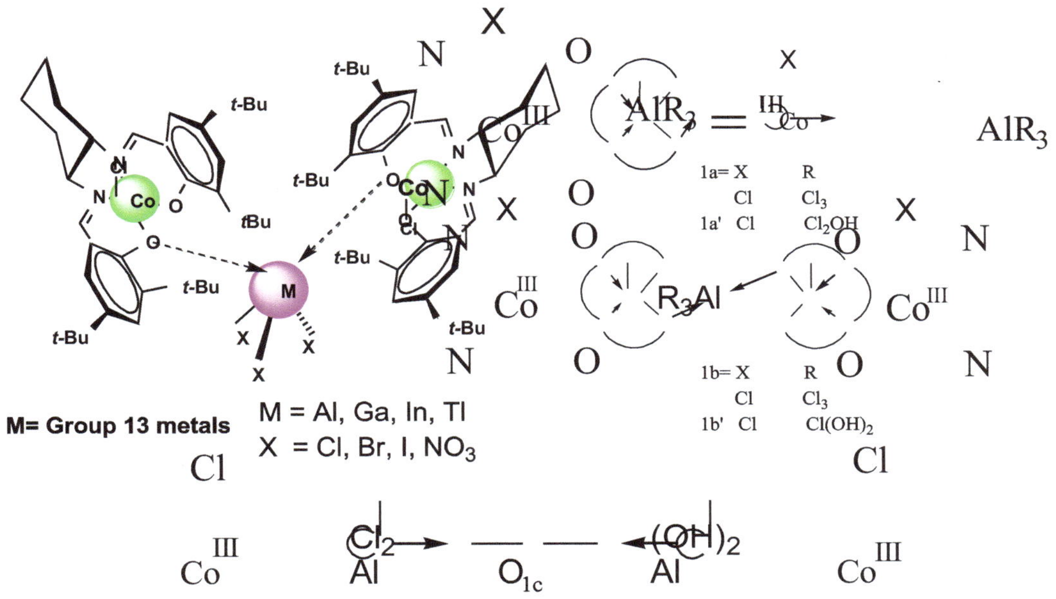

Scheme 9. Mono-and Bimetallic Chiral (Salen)Co Complexes.

Thakur and Kim *et al.* [34 - 38] developed highly active and enantioselective bim-etallic chiral (salen) Co catalyst by activating the inactive (salen)Co^{II} pre-catalyst by solid Lewis acids of group 13 metal salts (Scheme **9**). These bimetallic complexes showed very efficient in HKR of terminal epoxides and other asymmetric ring-opening reactions. The bimetallic [(salen)Co]$_2$Al complex display high enantioselectivity (95 – >99 ee %) and reactivity (within 6 hours) in the HKR of diverse terminal epoxides even in low loading of catalyst (0.5 mol %) with respect to their monometallic complex (Fig. **1**) [34].

The other bimetallic catalysts (Co-Ga, Co-In, Co-Tl) provided similar recovered high enantioselectivity (98 – >99 ee %) and reactivity, in the HKR of diverse terminal epoxides in low loading of catalyst (0.5 mol %) with respect to their monometallic complex (Scheme **10**) [35 - 37].

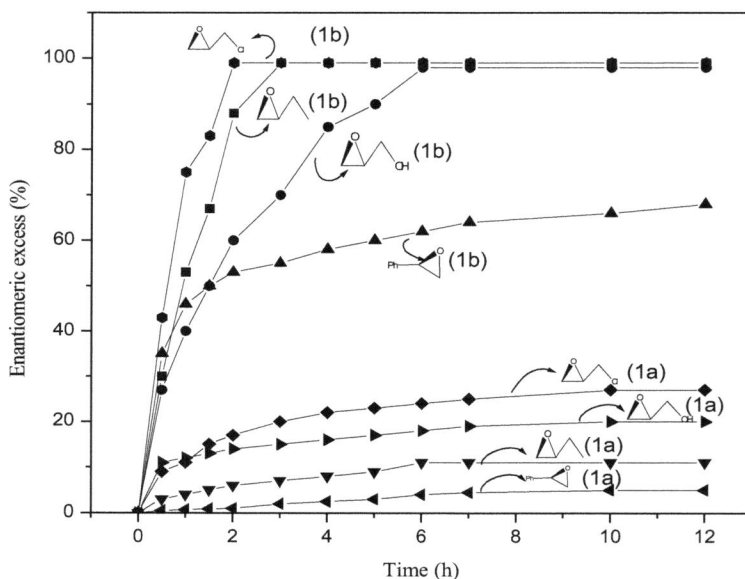

Fig. (1). The catalytic activity of monometallic and bimetallic for the asymmetric HKR of terminal epoxides using a 0.5 mol % catalyst at ambient temp.

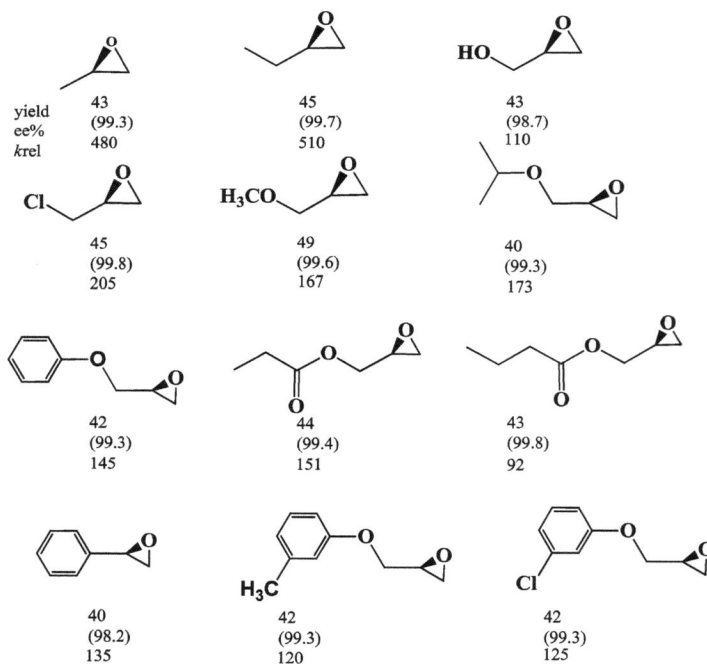

yield	43	45	43
ee%	(99.3)	(99.7)	(98.7)
krel	480	510	110

Scheme 10.

The corresponding epoxide products obtained after HKR are vicinal diols or 1, 2-diols which showed > 82 to > 89% *ee* except for glycerol (since it is achiral in nature) and > 32 to 44% isolated yield (Scheme **11**). Thus, these chiral vicinal diols are valuable intermediates and auxiliaries in organic and pharmaceutical synthesis.

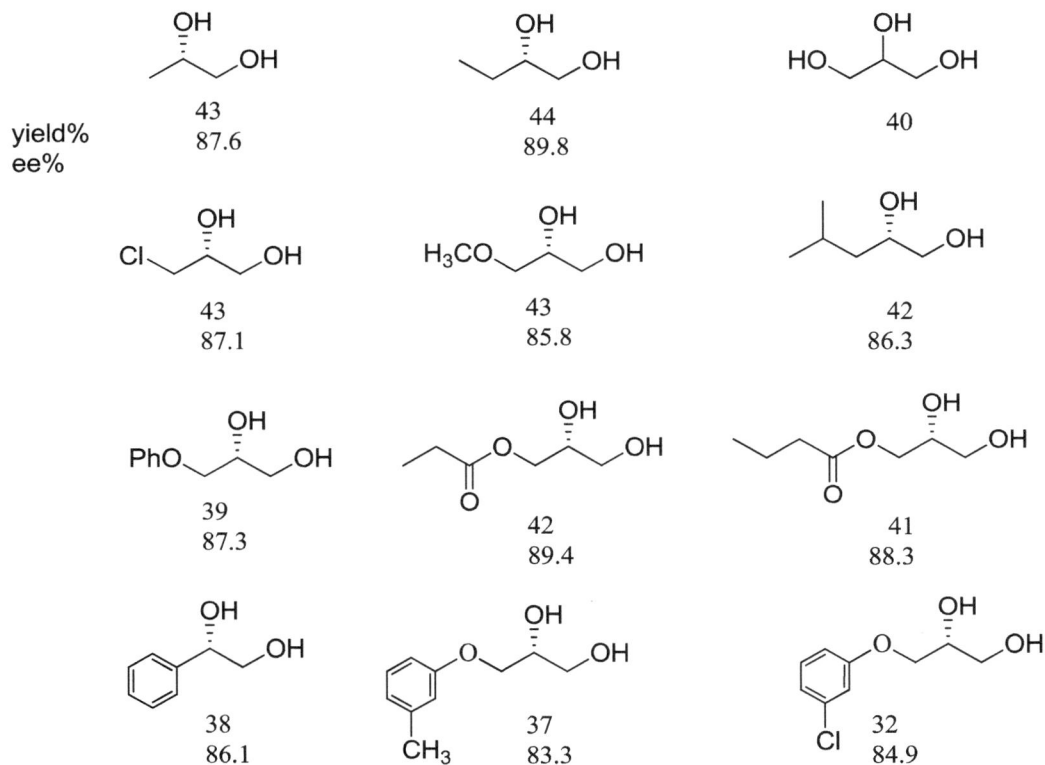

Scheme 11.

The homo- and heterogeneous bimetallic chiral Co(salen) complexes linked with group 13 metals have been showed high catalytic reactivity and enantioselectivity not only for the asymmetric ring-opening of epoxides with water, chloride ions, and carboxylic acids but also catalyzes the asymmetric cyclization of the ring-opened products, to prepare optically pure terminal epoxides in a single step (Scheme **12**) [37b].

To explore the mechanistic pathway, the kinetic studies (Table **3**) of the HKR of epichlorohydrin (ECH) have been done, which shows the two-term rate equation involving both intra and intermolecular components (Eq 2) [13, 14]. Plots of rate/[catalyst] *vs.* [catalyst] should be linear with slopes equal to k_{inter} and y-intercepts corresponding to k_{intra} catalysts 1b, 1b' and 1c revealed linear correlations with positive slopes and nonzero y-intercepts, consistent with the participation of both inter- and intra-molecular pathways in the HKR. A similar analysis of rate data obtained with monometallic catalysts 1a and 1a' revealed y-intercepts of zero, reflecting the absence of any first-order pathway for these catalysts. Thus, the dimer catalysts provide appropriate relative proximity and orientation which eventually reinforces the reactivity and selectivity relative to the monometallic complex [34].

$$\text{Rate } \alpha \ k_{\text{intra}} \ [\text{catalyst}] + k_{\text{inter}}[\text{catalyst}]^2 \qquad\qquad (2)$$

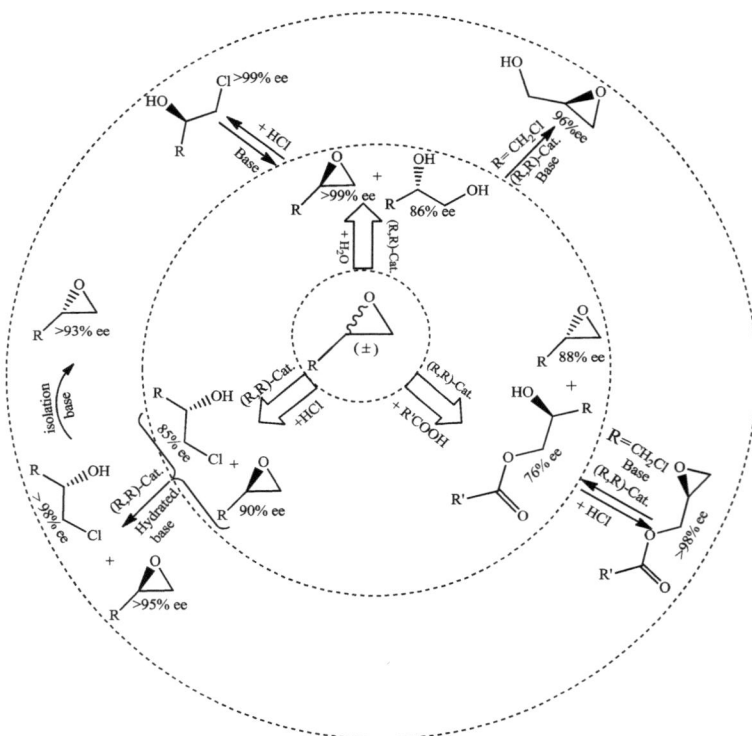

Scheme 12.

Table 3. Kinetic data for the HKR of racemic epichlorohydrin catalyzed by the monometallic and bimetallic [Co (salen):AlCl₃] catalysts.

Catalyst	No of (salen) Co Unit	k_{intra} (min⁻¹x 10⁻²)	k_{inter}(M⁻¹x min⁻¹)
Monometallic-1a	1	-	1.0
Monometallic-1a'	1		1.1

(Table 3) cont.....

Catalyst	No of (salen) Co Unit	k_{intra} (min^{-1}x 10^{-2})	k_{inter}(M^{-1}x min^{-1})
Bimetallic-1b	2	44.4	10.2
Bimetallic -1b'	2	45.9	15.7
Bimetallic -1c	2	48.9	17.7

The asymmetric ring-opening reactions follow the cooperative bimetallic mechanism proposed by Jacobsen *et al*. Thus, the bimetallic catalysts provide appropriate relative proximity and orientation which eventually reinforces the reactivity and selectivity relative to monometallic complex.

The HKR reactions follow the cooperative bimetallic catalysis where epoxide and nucleophile activate simultaneously by two different (salen)Co-Al or Ga catalyst molecules. The active intermediate during HKR may be a Co^{III}-OH complex [8a]. The linking of two (salen)Co unit through the Al and Ga induces the cooperative mechanism, albeit through a far less enantio-discriminating transition state than that attained with their monometallic catalyst. Unlikely to the dinuclear complex of Co-Al and Co-Ga, the monomers of Co-In and Co-Tl show the intramolecular pathway, which is quite evident from the kinetic analysis. It seems, Co-In and Co-Tl act as heterometallic complexes exhibiting two different Lewis acid centers Co and In and Co and Tl and show strong synergistic effect [39]. However, the dinuclear complex of Co-In and Co-Tl shows twofold more reactive than a corresponding monomeric analogy. In the catalytic cycle shown in Scheme **13**, the central metal ion Co activates and controls the orientation of incoming epoxide where only one enantiomer of it binds stereoselectively to the cobalt ion. The other metal ion (In or Tl), activates and binds the nucleophile H$_2$O, and as a result of these, enantioselective ring opening reactions of terminal epoxides takes place [37]. The proposed model for HKR catalyzed by Co-In and Co-Tl complex is given in Scheme **13** which may be similar to the enantioselective ring-opening of epoxides with 4- methoxyphenol catalyzed by gallium heterobimetallic complexes as reported by Shibasaki *et al*. [40]. The intra and intermolecular mechanisms for these catalysts may be similar to the earlier report [28b, 41, 42].

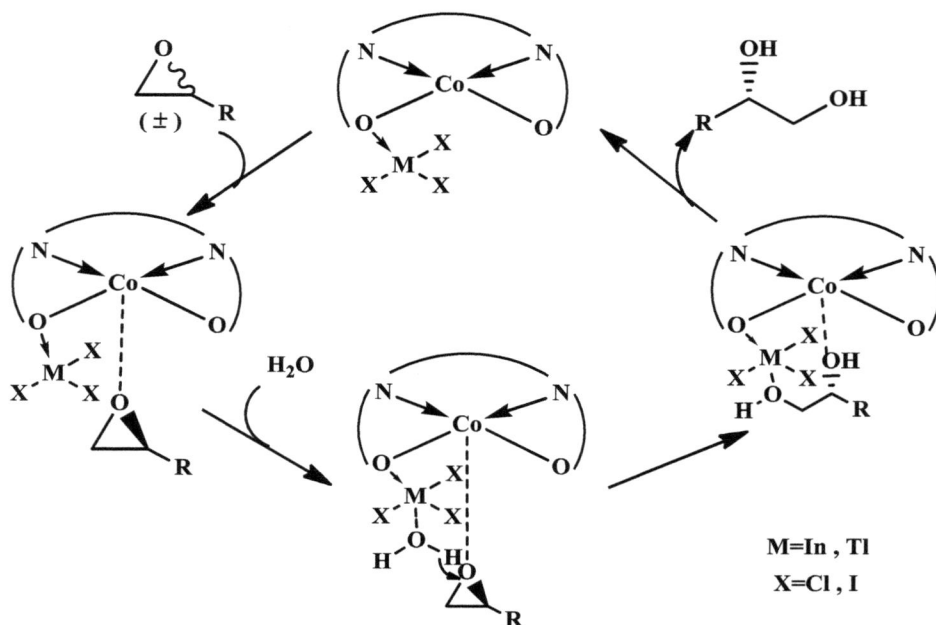

Scheme 13.

Quite recently, Thakur *et al.* [43, 44] utilized water-resistant Lewis acids like Lanthanide triflates to activate inactive chiral salen Co (II) compound to form bimetallic and monometallic corresponding Co (III) activated complexes. These complexes applied as chiral catalysts in HKR reactions of various terminal epoxides (Scheme **14**).

Kim and Thakur have reported the heterogenization of the chiral (Salen) Co [III] complexes on inorganic support having large surface area mesoporous material like siliceous MCM-41 through multi-step anchoring (Scheme **15**) and used it as catalysts in the hydrolytic kinetic resolution of racemic epoxides to diols [37b, 45, 46]. These bimetallic homo-and heterogeneous chiral (salen) Co catalysts have also been used in the various asymmetric ring-opening of terminal epoxides with Cl⁻ ions and carboxylic acids.

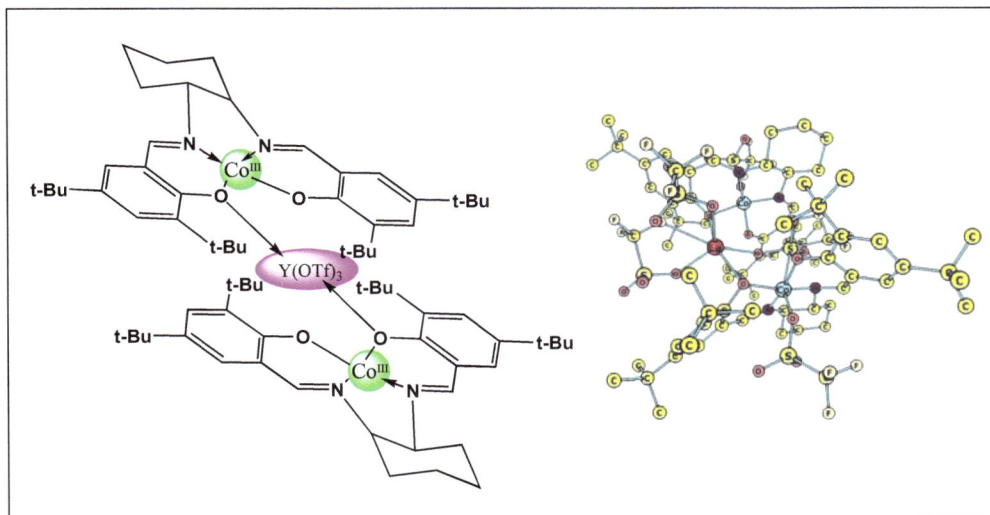

Scheme 14.

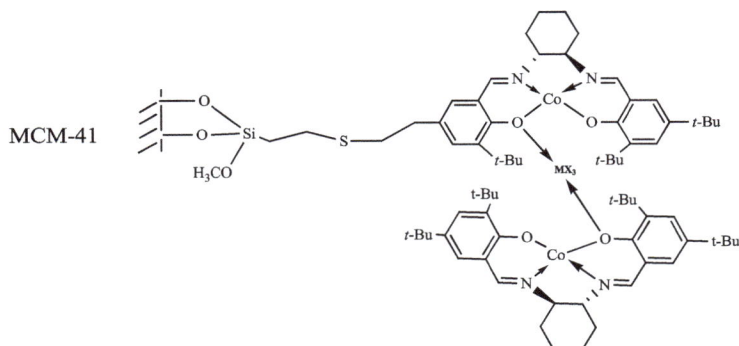

Scheme 15.

Jones *et al.* reported styryl substituted unsymmetrical "bisalen" and demonstrated its versatility in creating highly active, cooperative homogeneous and recyclable heterogeneous catalysts in the hydrolytic kinetic resolution of epoxides. With this catalyst (Scheme **16**) reactions were performed at a 0.02 mol% Co loading. At this concentration, the styryl unsymmetrical bisalen was highly active, achieving the full conversion of one enantiomer (99% ee, conversion 51%) in 7 h [47]. Using density functional theory (DFT) they have proposed the transition state structure, which is similar to the reported by Jacobsen *et al.* [48].

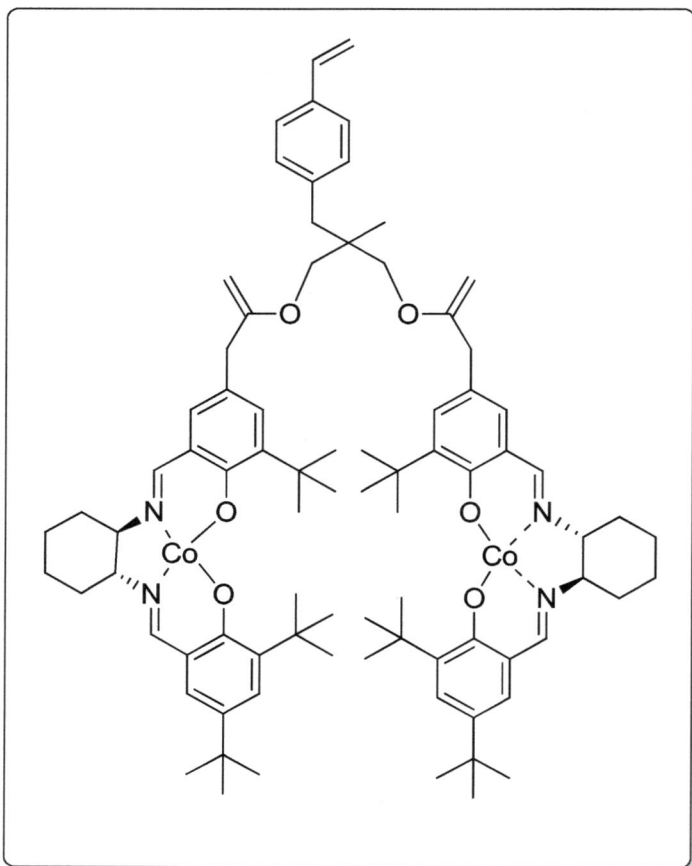

Scheme 16.

Kahn and Weck *et. al.* developed highly crosslinked polycyclooctyl chiral salen Co (III) and poly(styrene) resin-supported Co (III) salen cyclic recyclable oligomers (Scheme **17a**) showing significantly greater yield in the HKR of terminal epoxides and making this protocol synthetically more viable [49].

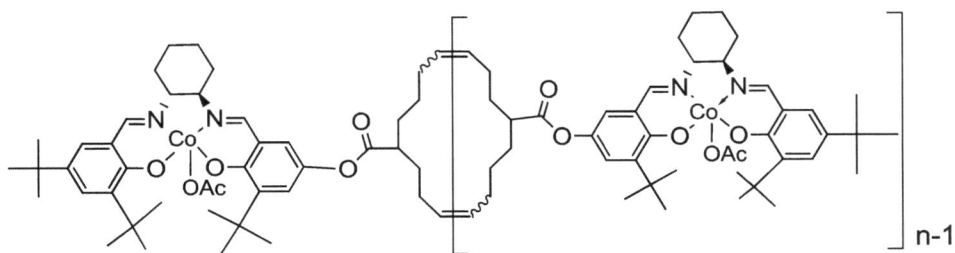

Scheme 17a.

The Co-salen complex is supported on macrocycles formed by the ring-expanding olefin metathesis of cyclooctene salen monomers using Grubb's 3[rd]generation initiator [50]. Unfortunately, the product distribution of the metathesis reaction produced mainly dimers and trimers, which were determined to be less or not at all active [51]. The multiple-step complicated synthetic procedure is another major problem.

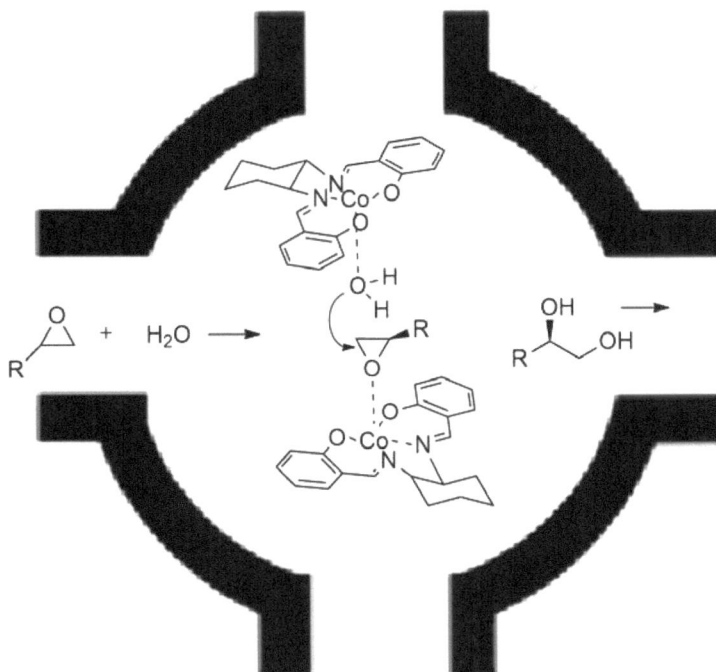

Scheme 17b.

Can Li research group [52] demonstrated the cooperative activation effect in the hydrolytic kinetic resolution (HKR) of epoxides by confining [Co(salen)] complexes in the isolated nanocages (pore size-tunable from 4-8 nm) of mesoporous silica SBA-16 (Scheme **17b**) with a high local concentration and thus generated a more active solid catalyst compared with the homogeneous counterpart of Jacobsen catalyst. However, enantioselectivity is not comparable with the homogeneous analogy.

Weberskirch group reported the self-assembled nanoreactors as highly active catalysts in the hydrolytic kinetic resolution (HKR) of epoxides in water using a (salen) Co^{III}complex supported on an amphiphilic, water-soluble block copolymer (Scheme **18**). However, generating nanoreactors, diffusion of reacting partners, and characterization of the catalyst is also a challenging work [53].

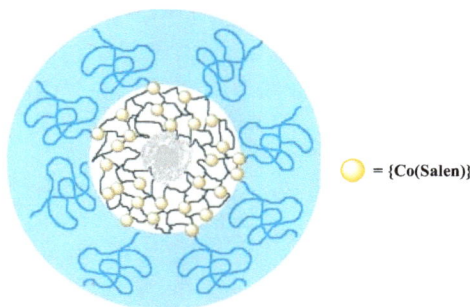

○ = {Co(Salen)}

Scheme 18.

Song *et al.* reported [54] that chiral (salen)Co^{II}supported on ionic liquid-based on 1-butyl-3-methylimidazolium [bmim]salt, is oxidized without Brønsted acids to catalytically active (salen)Co^{III} complex during HKR of racemic terminal epoxides (Scheme **19**). Further, they used the oxidatively pure (salen)Co^{III} complex in this reaction [55]. However, using the expensive ionic liquid makes this process less attractive strategy for the kinetic resolution [56].

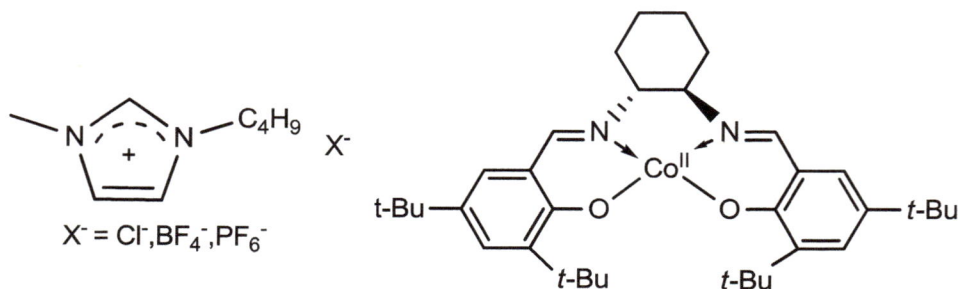

X⁻ = Cl⁻,BF$_4^-$,PF$_6^-$

Scheme 19.

Aerts *et al.* recycled the Jacobsen HKR homogeneous Co-Jacobsen complex in a semi-continuous operation using a laboratory prepared polymeric solvent resistant nano-filtration membrane [57]. They also measured the influence of solvent choice, acid activation, and surfactant addition on the hydrolytic kinetic resolution (HKR) of terminal epoxides [58]. This works deals with the recycling of the catalyst and reaction parameters of the HKR reaction.

Berkessel and Ertürk reported the HKR of epoxides catalyzed by chromium(III)-*endo,endo*-2,5-diaminonorbornane-salen   [Cr(III)-DIANAN--salen] complexes [59]. They claimed to have better activity and enantioselectivity than Jacobsen catalyst and reported, for the first time, non-enzymatic ring-opening of 2,2-disubstituted epoxides (*e.g.* 2-methyl-2-*n*-pentyloxirane) with water using [Cr(III)-DIANANE-salen complexes (Scheme **20**).

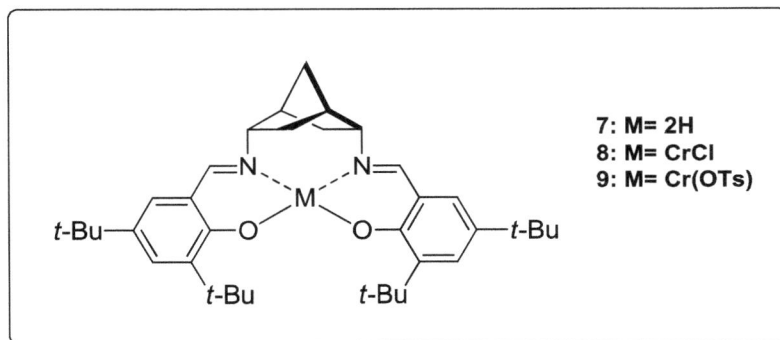

Scheme 20.

Beigi *et al.* reported polyglycerol-supported Co- and Mn- salen complexes as efficient and recyclable homogeneous catalysts for the hydrolytic kinetic resolution of terminal epoxides and asymmetric olefin epoxidation and observed the kinetic studies for HKR of 1, 2-epoxyhexane applying polyglycerol-supported Co-salen (Scheme **21**), which displayed enhanced catalytic activity compared to the respective non-immobilized catalyst [60].

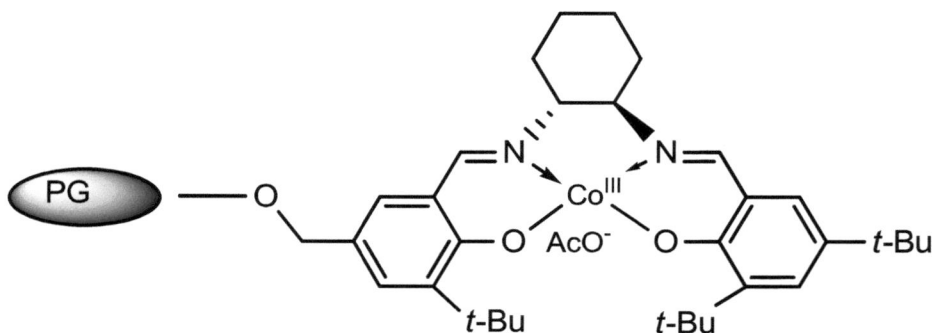

Scheme 21.

Wezenberg *et.al* [61]. prepared chiral, bimetallic cobalt (III) salen-calix [4]arene hybrid by using the upper rim of the calixarene scaffold and used as a chiral catalyst in the hydrolytic kinetic resolution (HKR) of three racemic epoxides. Kinetic studies on the HKR of *rac* 1,2-epoxyhexane revealed reaction follows an intra-molecular, cooperative pathway with k_{intra} 3.32 min^{-1} which gives 97.1% ee (52% conversion) within 8 hours by taking slightly excess of water 0.55 and 0.3 mol% of the bimetallic catalyst loading. Further, it is proposed that higher loadings of Co(III) sites on the same scaffold create a better statistical probability for the formation of cooperative bimetallic transition state geometry in such a way that, it ultimately enhances the rate of reaction.

Park *et al.* synthesized and utilized bis-urea-functionalized (salen)cobalt complexes (Scheme **22**) for the hydrolytic kinetic resolution of epoxides applying a self-assembly approach toward chiral bimetallic catalysts [62]. The complexes were synthesized to form self-assembled structures in solution through intermolecular urea–urea hydrogen-bonding interactions. These bis-urea (salen)Co catalysts displayed in rate acceleration (up to 13 times) in the hydrolytic kinetic resolution (HKR) of *rac*-epichlorohydrin in THF solvent by facilitating cooperative activation, in comparison to the monomeric catalyst. Further, the bis-urea (salen)CoIII catalyst 1k.OTs (Scheme **22**) very efficiently resolves butylene terminal epoxide (ee=99% and yield =43%) even under solvent-free conditions by requiring much shorter reaction time (8h) at low catalyst loading (0.03 mol%). For getting the same result the monomeric counterpart 2.OTS requires longer reaction time (24 h). It has been concluded that self-assembly through urea–urea hydrogen bonding is responsible for the observed rate enhancement over the covalently linked multimetallic catalysts.

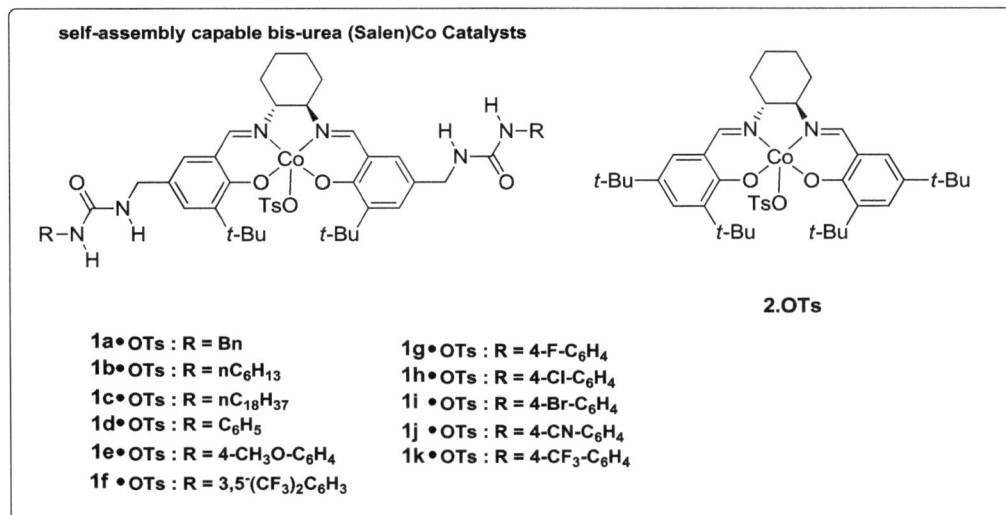

self-assembly capable bis-urea (Salen)Co Catalysts

2.OTs

1a•OTs : R = Bn
1b•OTs : R = nC$_6$H$_{13}$
1c•OTs : R = nC$_{18}$H$_{37}$
1d•OTs : R = C$_6$H$_5$
1e•OTs : R = 4-CH$_3$O-C$_6$H$_4$
1f •OTs : R = 3,5'(CF$_3$)$_2$C$_6$H$_3$

1g•OTs : R = 4-F-C$_6$H$_4$
1h•OTs : R = 4-Cl-C$_6$H$_4$
1i •OTs : R = 4-Br-C$_6$H$_4$
1j •OTs : R = 4-CN-C$_6$H$_4$
1k•OTs : R = 4-CF$_3$-C$_6$H$_4$

Scheme 22.

Cui research group developed chiral nanoporous metal−organic frameworks (MOF) are constructed by using dicarboxyl-functionalized chiral Ni(salen) and Co(salen) ligands. The Co(salen)-based framework is shown to be an efficient and recyclable heterogeneous catalyst for hydrolytic kinetic resolution (HKR) of racemic epoxides with up to 99.5% ee. It has been stated that MOF structure (Scheme **23**) organizes Co(salen) units into a highly dense arrangement and close proximity that improved bimetallic cooperative interactions, leading to very efficient catalytic activity in terms of enantiomeric excess and yield in HKR compared with its homogeneous analogs when low catalyst/substrate ratios are taken [63].

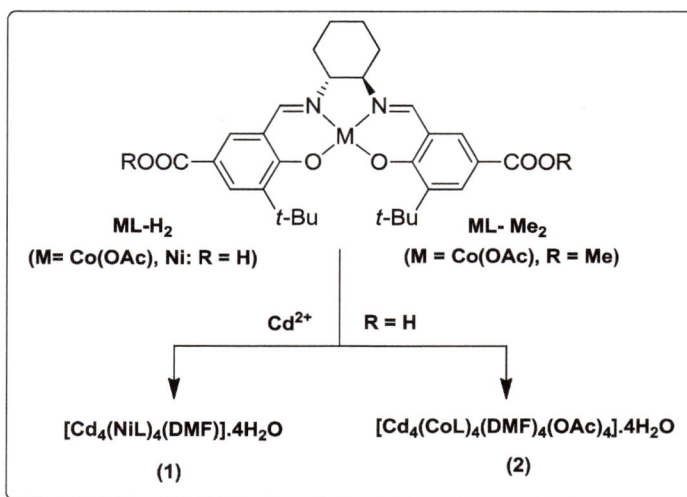

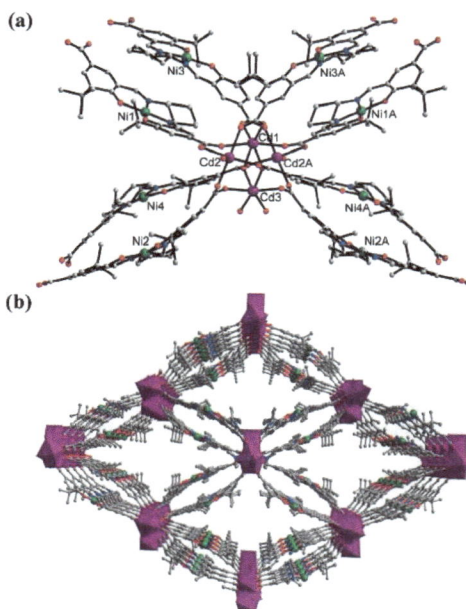

Scheme 23.

Kurahashi and Fujii explained the unique ligand-radical character of an activated cobalt salen catalyst that is generated by aerobic oxidation of a cobalt (II) salen complex and investigated that electronic structure of the Co(salen)(X) catalyst contains both Co^{III}(salen)(X) and Co^{II}(salen•+)(X) character (Scheme 24), in contrast to the conventional assignment as a cobalt(III) complex. Such a unique electronic structure of Co(salen)(X) is distinct from Cr^{III}-, Mn^{III}-, and Fe^{III}(salen)(X) complexes [64].

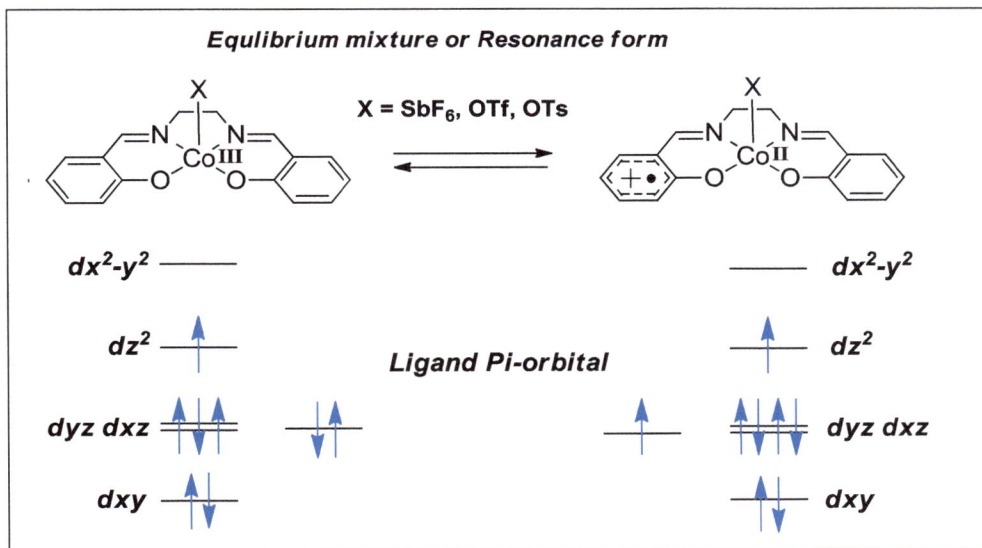

Scheme 24.

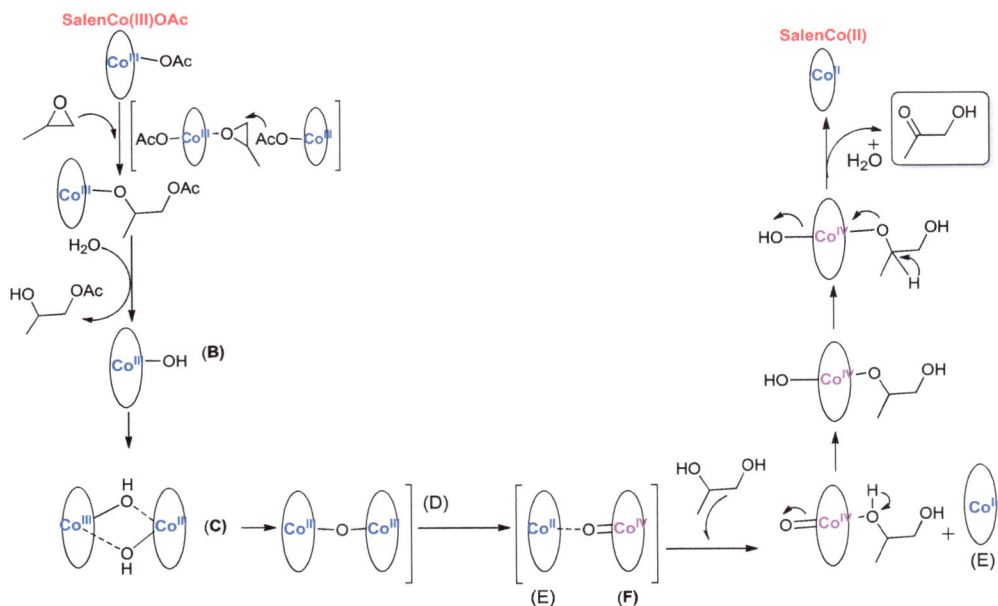

Scheme 25.

Ren *et al.* explored mechanistic aspects of change of oxidation of Co (III) to Co(II) in salenCo(III)OAc-catalyzed hydrolytic kinetic resolution of racemic epoxides [65]. It has been observed that the intermediate salenCo(III)OH plays an important role in the HKR process concerning the formation of a dimer (C) in (Scheme 25). Highly reactive oxo-cobalt (IV) is suggested to be formed by dehydration and disproportionation of the dimer based on ESI-MS investigation and is responsible for the oxidation of diols to afford α-hydroxy ketones.

Hong *et al.* developed chemical-promoted oxidative polymerization of modified-cobalt salen complexes (Scheme 26) used as efficient catalysts for the dynamic HKR of epibromohydrin with a maximum 80% yield and 86% ee at 48 h using 2 mol% of catalyst in THF [66]. In conventional kinetic resolution, maximum 50% theoretical yield can be obtained, whereas in case of dynamic kinetic resolution (DKR) 100% theoretical yield is possible for desired isomer. To achieve this goal, the substrate itself has to undergo racemization, and in this way the non-reacting enantiomer is transformed into the reacting one. The requisite condition for the DKR to be efficient, the rate constant for racemisation k_{rac} has to be equal to or greater than the rate constant k_F (rate constant for forward reaction) for converting substrate to product.

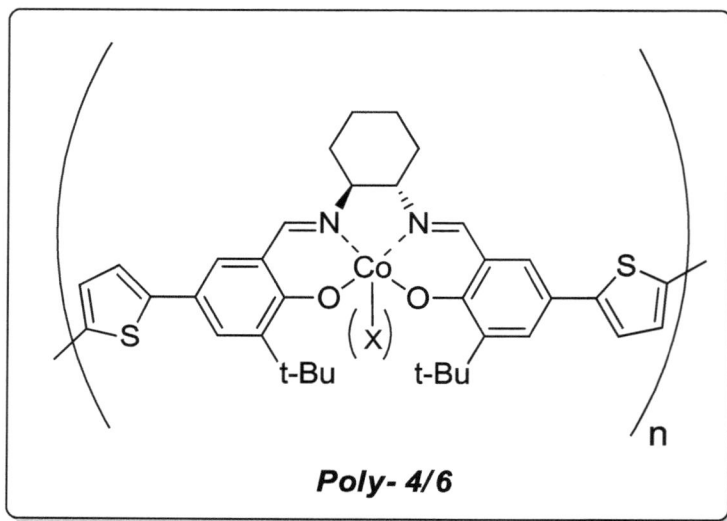

Poly- 4/6

Scheme 26.

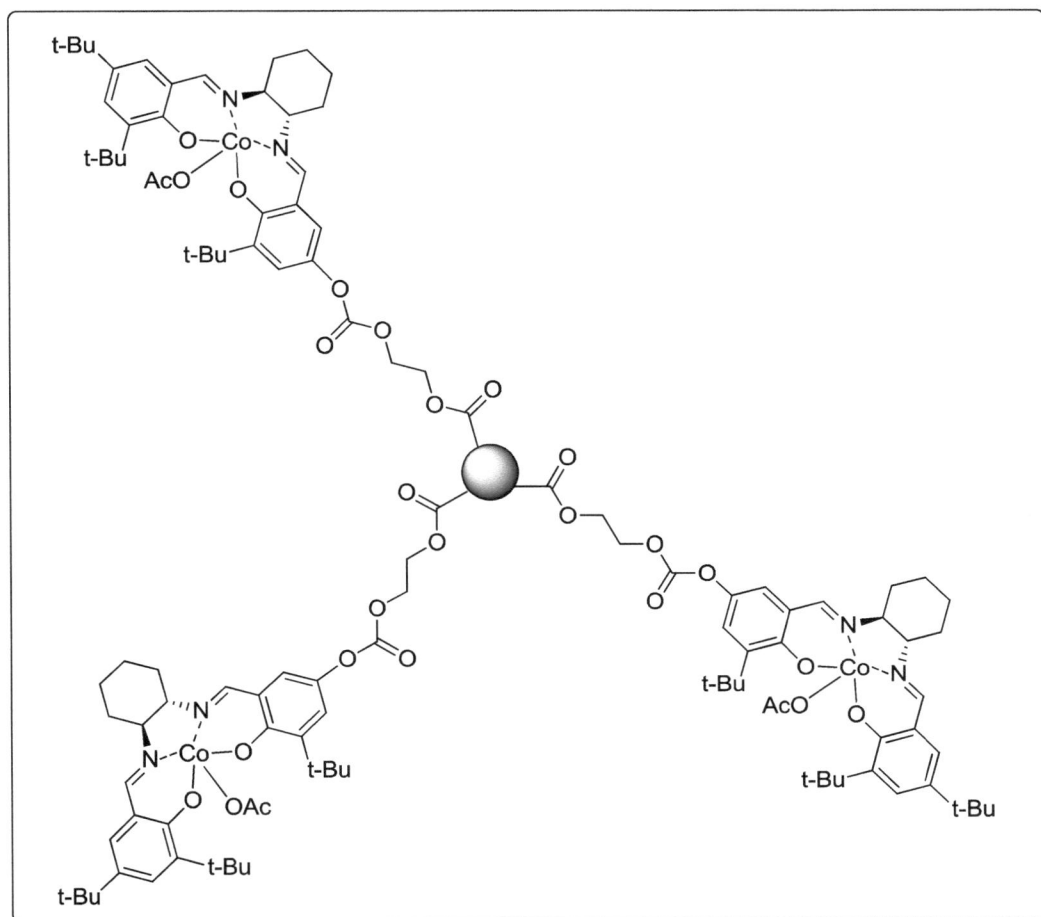

Scheme 27.

Matkiewicz *et al.* developed a series of novel polymer-supported chiral Co(III)–salen catalysts based on the hydroxyl functionalized gel-type resins, the 2-hydroxyethyl methacrylate (HEMA) resins. In the presence of this catalyst (Scheme **27**), the HKR of epichlorohydrin could be performed even at 0.01 mol% concentration of Co(III) ions resulting >99 ee% (conversion%= 52) in 27 h. The catalysts could be recycled ten times at 0.5 mol % concentration without any additional reactivation [67].

Bredihhina *et al.* explored hydrolytic and aminolytic kinetic resolution of terminal bisepoxides catalyzed by (salen)Co^{III} complexes exhibited epoxy-diols and N-protected epoxy-amino alcohols with excellent enantio- and diastereoselectivity and good yields (Scheme **28**) [68].

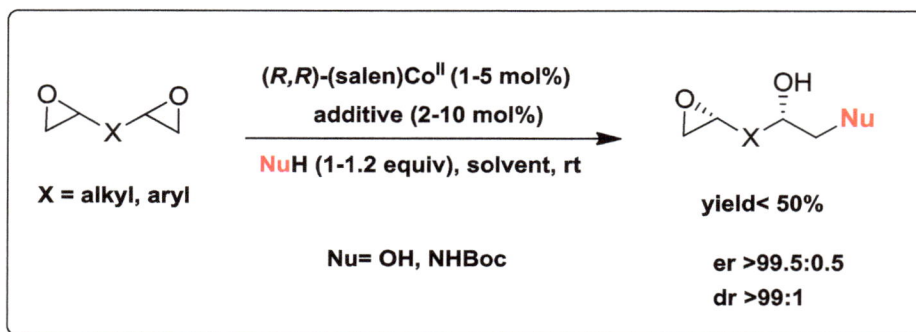

Scheme 28.

Wu *et al.* synthesized three binuclear Co(III)-salen complexes (Scheme **29**) based on a series of hydrophilic cyclic frameworks with different ring sizes, and superior catalytic performance have been observed with respect to monomeric counterpart, in the HKR reactions of terminal epoxides [69]. As the ring size (n) increases, the catalytic performances also increase in this process.

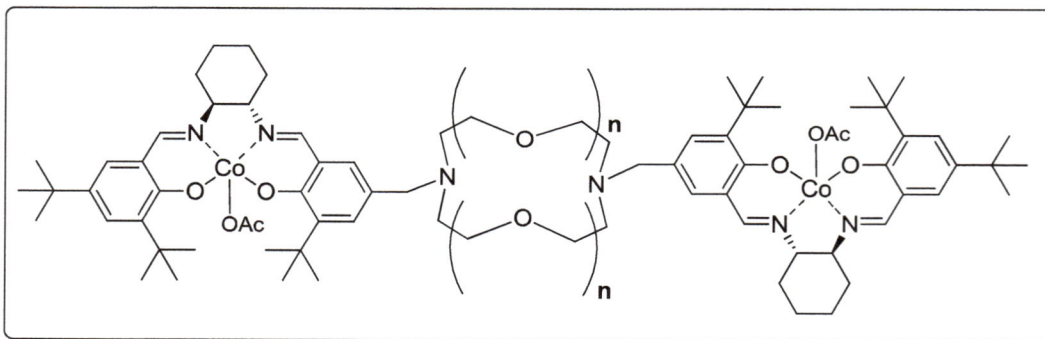

Scheme 29.

Sherrill research group investigated counter ion and substrate effects on barrier heights of the hydrolytic kinetic resolution of terminal epoxides (Scheme **30**) catalyzed by co(III)-salen using density functional theory (DFT) [70]. It was concluded that the bicatalyzed reactions show a non-additive, cooperative catalysis effect with a greater barrier height reduction than one would expect by summing the barrier height reductions from activation of the epoxide and activation of the nucleophile.

It has also been found epichlorohydrin in the HKR transition state and calculated that its barrier height was approximately half (18.8 kJ mol^{-1}), as large as those of 1-hexene oxide and propylene oxide (33.2 and 35.5 kJ mol^{-1} respectively), which is qualitatively consistent with epichlorohydrin reacting faster experimentally.

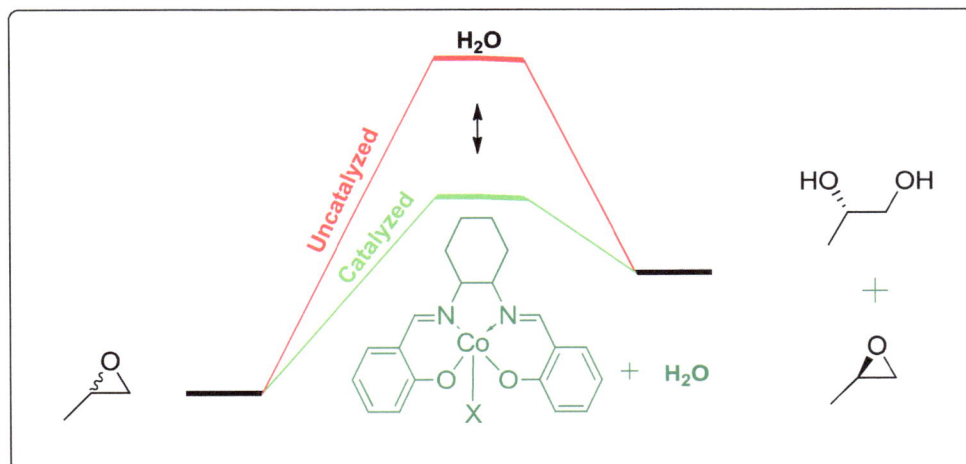

Scheme 30.

Bukowska *et al.* reported 2-hydroxyethyl methacrylate (HEMA) resin supported chiral salen Co (III) (Scheme **31**) and applied it in the hydrolytic kinetic resolution of racemic glycidyl esters which afforded 96% ee with good yield [71].

Dandachi *et al.* reported the mixing and matching of chiral cobalt- and manganese-based heterogenized and recyclable Calix-salen catalysts for the asymmetric hydrolytic ring opening of meso epoxides and DKR of epibromohydrin [72]. They prepared by using homochiral oligomeric salen macrocycles possessing aromatic spacer's new Calix-salen derivatives.

Kureshy *et al.* developed a homochiral bimetallic Co(III) Schiff base complex (Scheme **32**) and utilized it in the HKR of epichlorohydrin, propene oxide, and styrene oxide with low loading amount of catalyst (0.3 mol%) [73]. The catalytic performance in the HKR of epichlorohydrin was measured by the dimeric Co(III).BF$_4$ complex (BF$_4$- is taken as a counter ion in place of -OAc in Scheme **32**) mixed with Co(II) salen Jacobsen's catalyst (molar ratio 1:2) and got the turn over frequency (TOF) of this catalyst equal to the 31.96 (h^{-1}). This increase in TOF and performance is consistent with the polymeric Co–salen complex reported by Kim *et al.* [33a].

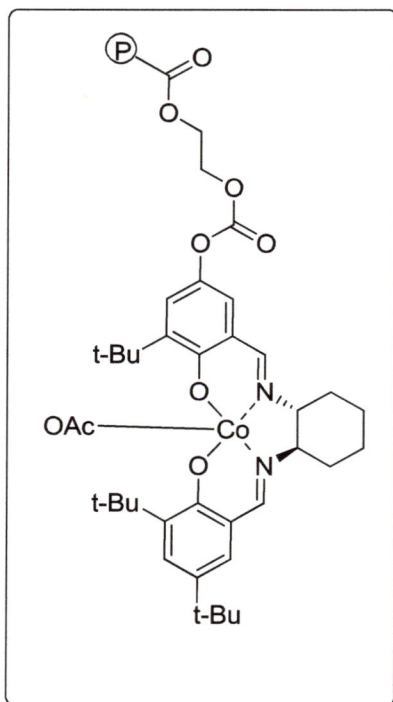

Scheme 31.

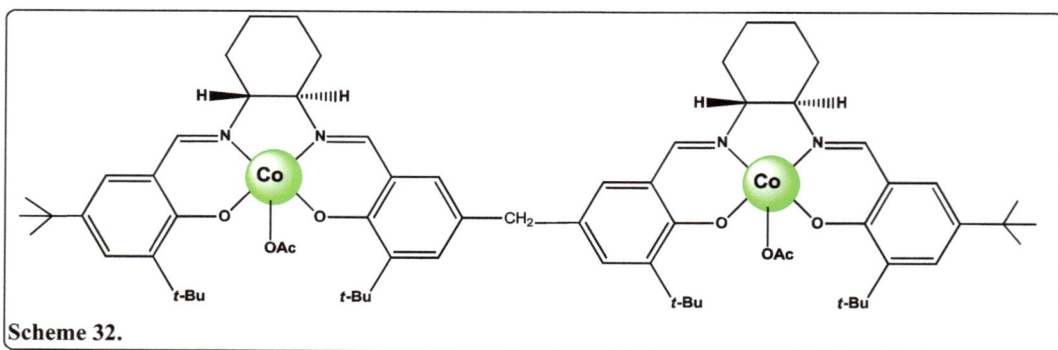

Scheme 32.

Very recently the same research group developed new chiral macrocyclic Co(III) salen complexes (Scheme **33**) and used as a catalyst in the asymmetric kinetic resolution of terminal epoxides and glycidyl ether using amines and water as nucleophiles and exerted 49% yield and ee upto 99% [74].

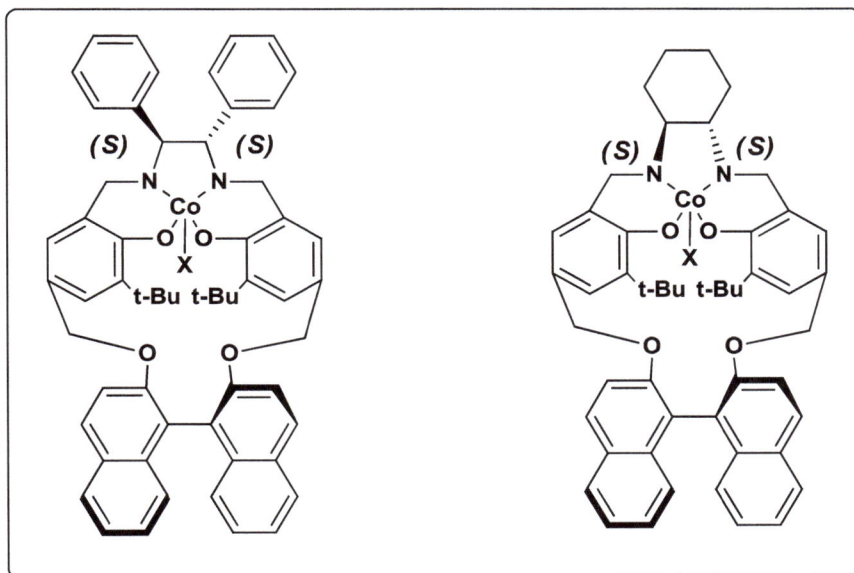

Scheme 33.

Liu *et al.* developed a ceramic membrane reactor having immobilized chiral di-chloromethyl salen-Co catalyst and performed HKR of racemic epichlorohydrin on to it (Fig. **2**) [75].

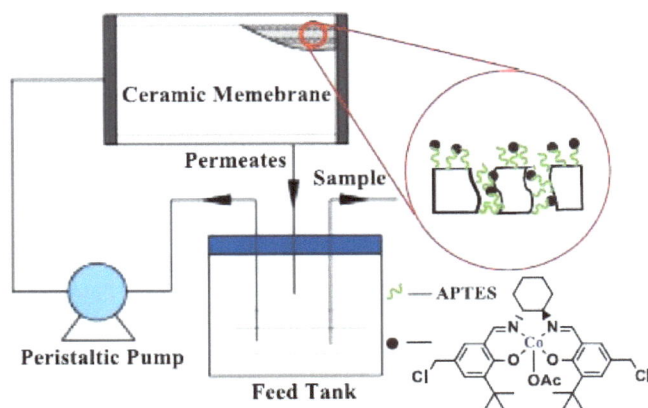

Fig. (2). Ceramic membrane reactor.

Very recently, Cozzi *et al.* research group developed air-stable compound *i.e.*

[Al(Salophen)] derivatives that showed spontaneous self-assembly at the graphite/solution interface by forming highly-ordered nano-patterns (Scheme **34**). These materials are having potential applications in material chemistry and catalysis [76].

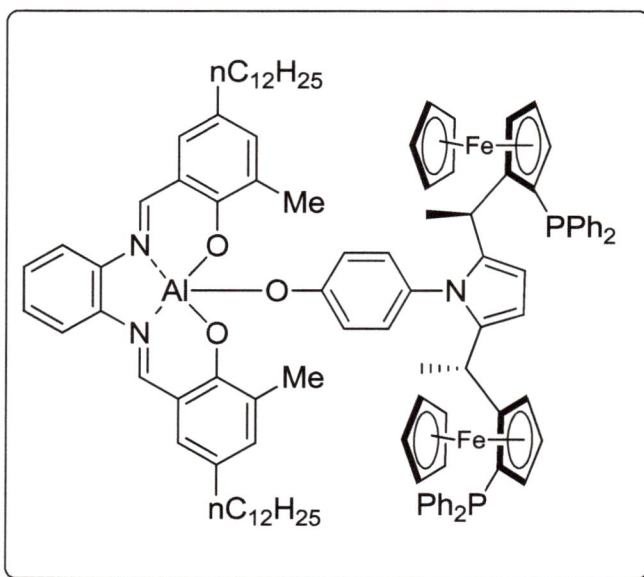

Scheme 34.

Zhang *et al.* successfully encapsulated homogeneous Co(salen) and Ti (salen) complexes by diffusion-limited atomic layer deposition (ALD) into the nano-channels of SBA-15, SBA-16, and MCM-41 (Scheme **35**) which shows excellent catalytic activity and reusability in the hydrolytic kinetic resolution of epoxides and asymmetric cyanosilylation of carbonyl compounds, respectively [77].

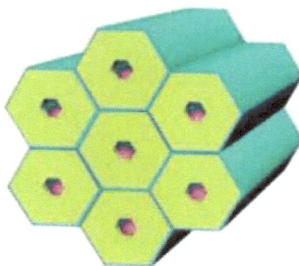

Scheme 35.

Mower and Blackmond very recently explored vibrational circular dichroism combined with FTIR spectroscopy (VCDIR) as an important analytical tool in order to get detailed analysis on complex mechanistic information of the Jacobsen hydrolytic kinetic resolution. The observed findings resemble experimental kinetic studies and computational investigations. This work promises to add for mechanistic understanding, reaction monitoring, and process improvement in asymmetric catalytic transformations [78].

For a better understanding of kinetics and mechanism of a hydrolytic kinetic resolution of terminal epoxides theoretical studies like density functional theory have been performed.

Li *et al.* [79] calculated DFT calculations by exchange-correlation interaction (B3LYP) method using a basis set of Dunning/Huzinaga valence double-zeta and were performed for a cooperative activation in ring-opening hydrolysis of propylene epoxide catalyzed by Co-salen (III) complex.

They showed that the activation energies of the reactions with two Co-salen catalysts are significantly lower (32 kJ/mol) in comparison to considering single Co-salen catalyst (97-174kJ/mol). In addition to the cooperative bimetallic mechanism, it also promotes the cooperative charge transfer during the reactions (Scheme **36**). The transition state analysis of an HKR reaction shows S_N^2 reaction pathway and second-order kinetics dependence on the concentration of the catalysts *i.e.* nucleophilic [(salen)Co-OH] and activated [(salen)Co-*R* or *S*-propylene epoxide present during the course of a chemical reaction. In contrast to the common S_N^2 reaction, in this case, there is no inversion of handedness since the nucleophilic OH preferentially attacks the primary achiral carbon atom of the propylene epoxide. These theoretical studies are close agreement with the experimental value.

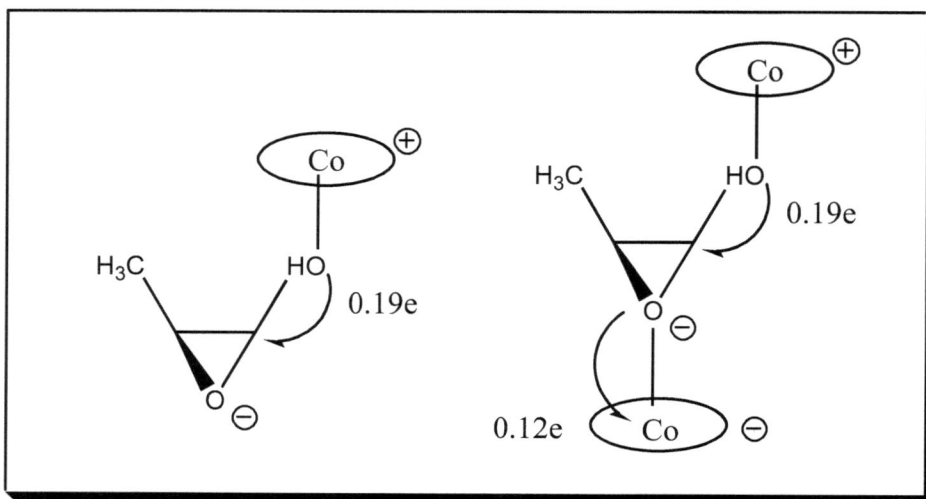

Scheme 36.

Jacobsen *et al.* made a computational calculation using DFT and B3LYP/6-31G(d) method for an HKR of propylene epoxide by (salen)Co (III) complex. It has been observed that in a cooperative bimetallic mechanism, the stereochemistry of each of the two (salen)Co(III) catalysts is very crucial in the rate-determining transition structure to form a product. Matching the absolute stereochemistry of each of these (salen)Co(III) catalysts (*R,R* or *S,S*) is a prerequisite which needs minimum activation energy to proceed for a measurable rate of hydrolysis (Scheme **37**). They further concluded on the basis of strong evidence of experimental and computational studies that in the HKR it is the stepped conformation of the salen ligand who plays a significant role in a stereochemical communication instead of the shape of chiral diamine backbone of the ligand [48].

Sherrill research group studied counter ion and substrate effects on reaction barrier heights of the hydrolytic kinetic resolution of terminal epoxides catalyzed by Co(III)-salen calculated computationally using density functional theory (DFT). They computed by taking three experimentally active counter ions *i.e.* chloride, acetate, and tosylate, as well as −OH, which is thought to be very important in the rate-limiting step and determined the activation energies as 35, 38, 34 and 48 kJ mol^{-1} respectively. The activation energies for propylene oxide, 1-hexene oxide, and epichlorohydrin are 35.5, 33.2 and 18.8 kJ mol^{-1}respectively when catalyzed by Co-salen-OAc showing faster reactions for epichlorohydrin.

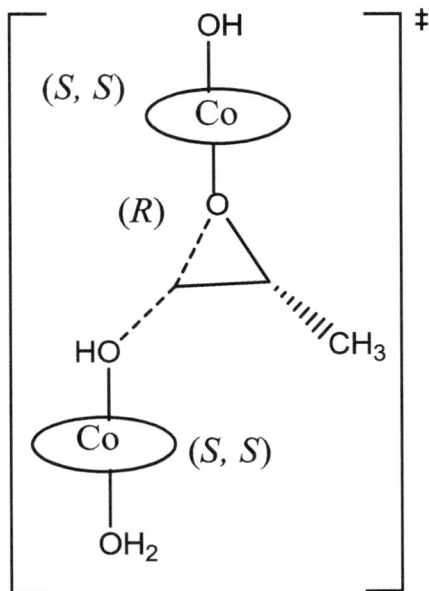

OH

(S, S)   Co

(R)   O

HO   CH₃

Co   (S, S)

OH₂

‡

Scheme 37.

CONCLUSION AND PROSPECTS

The bi-and multimetallic chiral Co- salen proved to be a very promising and efficient chiral catalyst in hydrolytic kinetic resolution (HKR) of racemic terminal epoxides for synthesizing several classes of chiral building blocks. The chiral salen ligand structural framework is synthetically flexible for tuning both sterically and electronically to prepare bi- and multimetallic complexes as asymmetric catalysts. In the HKR of terminal epoxides, the involvement of two or more chiral salen [Co] units not only enhances the reaction rates by intra-molecular cooperative bimetallic mechanism but also provides a general method to prepare many valuable chiral epoxides and 1, 2-diols in high yields, as well as enantio- purity in an efficient and greener way from cheaply available racemic substrates. The rate of enhancement of chiral salen Co bimetallic catalyst has been explained by kinetic studies of the HKR of terminal epoxides by showing the additional k_{intra} term, which is lacking in the monometallic catalyst. Most of the HKR reactions are performed in solvent-free conditions, which greatly assist the green chemistry philosophy in reducing/excluding undesired side products and hazardous materials. The HKR reactions and chiral building blocks obtained by HKR reactions have wide applications in the synthesis of bioactive compounds [80].

The chiral salen bimetallic complexes, in which two metals (same or different)

Lewis acid centers function cooperatively, has also been implicated in diverse valuable asymmetric transformations, such as asymmetric epoxide ring-opening reactions with diverse nucleophiles [81, 82], ring-opening of oxetanes [83], (salen)Al(III)-catalyzed conjugate addition reactions [84], stereoselective epoxide polymerization [85], and epoxide/CO_2 copolymerization [86], *etc.* The results presented in this chapter clearly demonstrate the need to design and develop a highly efficient novel chiral catalyst for asymmetric synthesis in order to meet the demanding requirements of optically valuable active substances for diverse applications. We believe research and development in the area of enantioselective kinetic resolutions are expected to have a promising future under the aegis of asymmetric catalytic synthesis. To explore the alternatives of the existing chiral organometallic catalysts for asymmetric catalytic synthesis, by simple and easily accessible organocatalysts (in a metal free condition) with high activity, selectivity, and recyclability remains an important challenge in this area.

CONSENT FOR PUBLICATION

Not applicable.

CONFLICT OF INTEREST

The author declares that there is no conflict of interest in this chapter.

ACKNOWLEDGEMENTS

The first (SST) and third author (NN) greatly acknowledge the Chhattisgarh Council of Science and Technology (CCOST), Raipur (C.G.) India for the financial support with a minor research project grant [No. 2236/CCOST/MRP/dated 23-12-2015] and Prof. Geon-Joong Kim, Inha University, South Korea for helpful discussion, for this work.

REFERENCES

[1]   Noyori, R. *Asymmetric Catalysis in Organic Synthesis*; Wiley: New York, **1994**. b) Jacobsen, E.N.; Pfaltz, A.; Yamamoto, H. *Comprehensive Asymmetric Catalysis I-III*; Springer Verlag: Berlin, **1999**.

[2]   Blaser, H.U.; Schmidt, E., Eds. *Asymmetric Catalysis on Industrial Scale: Challenges, Approaches and Solutions*; Wiley-VCH: Weinheim, **2004**. bCollins, A.N.; Sheldrake, G.N.; Crosby, J., Eds. *Chirality in Industry I and II*; John Wiley & Sons: Chichester, **1992 & 1996**. Sheldon, R.A. *Chirotechnology: Industrial Synthesis of Optically Active Compounds*; Marcel Dekker: New York, **1993**. Hanessian, S.; Delorme, D.; Tyler, P.C.; Demailly, G.; Chapleur, Y. Carbohydrates as "Chiral Templates" in Organic Synthesis - Target: Boromycin.*Current Trends in Organic Synthesis*; Nazaki, H., Ed.; Pergamon Press Oxford: New York, **1993**, pp. 205-219.

[3]   Seebach, D.; Hungerbühler, E. Synthesis of enantiomerically pure compounds (EPC-Synthesis) tartaric acid, an ideal source of chiral building blocks for synthesis. In: *Modern Synthetic Methods*; Scheffold, R., Ed.; Salle Sauerländer: Frankfurt, Aarau, **1980**, pp. 91-173. Vol. *2*.

[4]   a) Penne, J.S. *Chiral Auxiliaries and Ligands in Asymmetric Synthesis*; Wiley-Interscience

Publication: New York, **1995**. b) Paquette, L.A., Ed. *Chiral Reagents for Asymmetric Synthesis. Handbook of Reagents in Organic Synthesis*; John Wiley & Sons: NJ, USA, **2003**.

[5]   a) Gold, V.; Loening, K.L.; McNaught, A.D.; Sehmi, P. Compendium of Chemical Terminology: IUPAC Recommendations. *VIII Oxford/London/Edinburgh/Boston/Palo Alto/Melbourne: Blackwell Scientific Oxford*, **1987**.b) Ojima, I., Ed. *Catalytic Asymmetric Synthesis*, 2nd ed.; Wiley-VCH: New York, **2000**.
[http://dx.doi.org/10.1002/0471721506] c) Jacobsen, E.N.; Pfaltz, A.; Yamamoto, H., Eds. *Comprehensive Asymmetric Catalysis I-III*; Springer Verlag: Berlin, **1999**.
[http://dx.doi.org/10.1007/978-3-642-58571-5] d) Morrison, J.D., Ed. *Asymmetric Synthesis*; Academic Press: New York, **1985**. Vol. *5*.e) Bosnich, B., Ed. *Asymmetric Catalysis*; Martinus Nijhoff: Dordrecht, **1986**. f) Nogradi, M. *Stereoselective Synthesis*; VCH: Weinheim, **1987**. g) Kagan, H.B. Asymmetric Synthesis Using Organometallic Catalysts.*Comprehensive Organometallic Chemistry*; Wilkinson, G.; Stone, F.G.A.; Abel, E.W., Eds.; Pergmon Press: Oxford, **1982**, pp. 463-498. Vol. *8*.
[http://dx.doi.org/10.1016/B978-008046518-0.00112-4] h) Brunner, H. Enantioselective Synthesis of Organic Compounds with Optically Active Transition Metal Catalysts in Substoichiometric Quantities. In: *Topics in Stereochemistry*; Eliel, E.L.; Wilen, S.H., Eds.; Wiley-Interscience: New York, **1988**, pp. 129-247. Vol. *18*.
[http://dx.doi.org/10.1002/9780470147276.ch3] i) Brown, J.M.; Davies, S.G. Chemical asymmetric synthesis. *Nature*, **1989**, *342*, 631-636.
[http://dx.doi.org/10.1038/342631a0] j) Noyori, R. Chiral metal complexes as discriminating molecular catalysts. *Science*, **1990**, *248*(4960), 1194-1199.
[http://dx.doi.org/10.1126/science.248.4960.1194] [PMID: 17809904] k) Noyori, R.; Kitamura, M. Enantioselective catalysis with metal complexes. In: An overview.*Modern synthetic methods*; Scheffold, R., Ed.; Springr-Verlag: Berlin, Heidelberg, **1989**, pp. 115-198.

[6]   Thakur, S.S.; Lee, J.E.; Lee, S.H.; Kim, J.M.; Song, C.E. Heterogeneous enantioselective catalysis using inorganic supports. In: *Handbook of asymmetric heterogeneous catalysis,* 1st ed; Ding, K.; Uozumi, Y., Eds.; Wiley-VCH Verlag: Weinheim, Germany, **2008**, pp. 25-72.
[http://dx.doi.org/10.1002/9783527623013.ch2]

[7]   a) Halpem, J. Asymmetric catalytic hydrogenation: mechanism and origin of enantioselection. In: *Asymmetric synthesis*; Morrison, J.D., Ed.; Academic Press: New York, **1985**.
b) Brown, J.M. Directed homogeneous hydrogenation. *Angew. Chem. Int. Ed. Engl.,* **1987**, *26*, 190-203.
[http://dx.doi.org/10.1002/anie.198701901]

[8]   Yoon, T.P.; Jacobsen, E.N. Privileged chiral catalysts. *Science,* **2003**, *299*(5613), 1691-1693.
[http://dx.doi.org/10.1126/science.1083622] [PMID: 12637734]

[9]   Knowles, W.S. Asymmetric hydrogenations (Nobel lecture). *Angew Chem Int Ed,*   **1998-2002**, *41*http://nobelprize.org/

[10]   a) Tarbell, D.S.; Carman, R.M.; Chapman, D.D. The chemistry of Fumagillin. *J. Am. Chem. Soc.,* **1961**, *83*, 3096-3113.
[http://dx.doi.org/10.1021/ja01475a029] b) Sigg, H.P.; Weber, H.P. Isolation and structure elucidation of ovalicin. *Helv. Chim. Acta,* **1968**, *51*(6), 1395-1408.
[http://dx.doi.org/10.1002/hlca.19680510624] [PMID: 5680744] c) Nakajima, H.; Takase, S.; Terano, H.; Tanaka, H. New antitumor substances, FR901463, FR901464 and FR901465. III. Structures of FR901463, FR901464 and FR901465. *J. Antibiot. (Tokyo),* **1997**, *50*(1), 96-99.
[http://dx.doi.org/10.7164/antibiotics.50.96] [PMID: 9066774]

[11]   Winstein, S.; Henderson, R.B. *Heterocyclic Compounds*; Elderfield, R.C., Ed.; Wiley: New York, **1950**. Vol. *1*.
b) Rao, A.S.; Paknikar, S.K.; Kirtane, J.G. Recent advances in the preparation and synthetic applications of oxiranes. *Tetrahedron*, **1983**, *39*, 2323-2367.
[http://dx.doi.org/10.1016/S0040-4020(01)91961-1] c) Smith, JG *Synthetically useful reactions of epoxides synthesis*, **1984**, *8*, 629-656.

[12]   a) Larcheveque, M.; Henrot, S. Synthesis of new acylpseudopeptides analogous to n-acetylmuramyl dipeptide (MDP). *Tetrahedron,* **1990**, *46*, 4277-4282.b) March, P.D.; Figueredo, M.; Font, J.; Monsalvatje, M. Synthesis of (+)-Methyl (R,E)-6-benzyloxy-4-hydroxy-2-hexenoate and its mesylate derivative. *Synth. Commun.,* **1995**, *25*, 331-342.
[http://dx.doi.org/10.1080/00397919508011364] c) Adiyaman, M.; Khanapure, S.P.; Hwang, S.W.; Rokach, J. Regioncontrolled formation of iodohy dnns and expoxides from vic-diols. *Tetrahedron Lett.,* **1995**, *36*, 7367-7370.
[http://dx.doi.org/10.1016/0040-4039(95)01658-9]

[13]   Katsuki, T., Ed. *Asymmetric Oxidation Reactions: A Practical Approach in Chemistry*; Oxford University Press Inc: New York, **2001**.

[14]   Katsuki, T. *Comprehensive Asymmetric Catalysis*; Jacobsen, E.N.; Pfaltz, A.; Yamamoto, H., Eds.; Springer: New York, **1999**.
Rossiter, B.E. *Asymmetric Synthesis*; Morrison, J.D., Ed.; Academic Press: New York, **1985**. Vol. 5.
Johnson, R.A.; Sharpless, K.B. *Catalytic Asymmetric Synthesis*; Ojima, I., Ed.; VCH: New York, **1993**.
dKatsuki, T.; Martin, V.S. Asymmetric epoxidation of allylic alcohols: The katsuki–sharpless epoxidation reaction. *Org. React.,* **1996**, *48*, 1-299.
[http://dx.doi.org/10.1002/0471264180.or048.01] eKatsuki, T.; Sharpless, K.B. The first practical method for asymmetric epoxidation. *J. Am. Chem. Soc.,* **1980**, *102*, 5974-5976.
[http://dx.doi.org/10.1021/ja00538a077]

[15]   a) Jacobsen, E.N.; Wu, M.H. *Comprehensive asymmetric catalysis*; b) Jacobsen, E.N.; Pfaltz, A.; Yamamoto, H., Eds.; Springer: New York, **1999**.
[http://dx.doi.org/10.1007/978-3-642-58571-5] c) Katsuki, T. Catalytic asymmetric oxidations using optically active (salen)Manganese(III) complexes as catalysts. *Coord. Chem. Rev.,* **1995**, *140*, 189-214.
[http://dx.doi.org/10.1016/0010-8545(94)01124-T] d) Jacobsen, E.N. *Comprehensive Organometallic Chemistry II*; Wilkinson, G.; Stone, F.G.A.; Abel, E.W.; Hegedus, L.S., Eds.; Pergamon: New York, **1995**, pp. 1097-1135. Vol. *12*.
[http://dx.doi.org/10.1016/B978-008046519-7.00137-4] e) Jacobsen, E.N.; Zhang, W.; Muci, A.R.; Ecker, J.R.; Li, D. Highly enantioselective epoxidation catalysts derived from l,2-diaminocyclohexane. *J. Am. Chem. Soc.,* **1991**, *113*, 7064-7066.
[http://dx.doi.org/10.1021/ja00018a068]

[16]   a) Kolb, H.C.; Sharpless, K.B. A simplified procedure for the stereospecific transformation of 1,2-diols into epoxides. *Tetrahedron,* **1992**, *48*, 10515-10530. For asymmetric dihydroxylation routes, see:
[http://dx.doi.org/10.1016/S0040-4020(01)88349-6] b) Corey, E.J.; Link, J.O. A catalytic enantioselective synthesis of denopamine, a useful drug for congestive heart failure. *Tetrahedron Lett.,* **1991**, *56*, 442-444. c) Corey, E.J.; Helal, C.J. A catalytic enantioselective synthesis of chiral monosubstituted oxiranes. *Tetrahedron Lett.,* **1993**, *34*, 5227-5230.
[http://dx.doi.org/10.1016/S0040-4039(00)73959-1] d) Ramachandran, P.V.; Gong, B.; Brown, H.C. Chiral synthesis *via* organoboranes. 40. selective reductions. 55. A simple one-pot synthesis of the enantiomers of trifluoromethyloxirane. A general synthesis in high optical purities of alpha.-trifluoromethyl secondary alcohols *via* the ring-cleavage reactions of the epoxide. *J. Org. Chem.,* **1995**, *60*, 41-46.
[http://dx.doi.org/10.1021/jo00106a012] e) Kitamura, M.; Tokunaga, M.; Noyori, R. Asymmetric hydrogenation of beta-keto phosphonates: a practical way to fosfomycin. *J. Am. Chem. Soc.,* **1995**, *117*, 2931-2932.
[http://dx.doi.org/10.1021/ja00115a030] f) Wang, Z-X.; Tu, Y.; Frohn, M.; Zhang, J-R.; Shi, Y. An efficient catalytic asymmetric epoxidation method. *J. Am. Chem. Soc.,* **1997**, *119*, 11224-11235.
[http://dx.doi.org/10.1021/ja972272g] g) Zhu, Y.; Wang, Q.; Cornwall, R.G.; Shi, Y. Organocatalytic asymmetric epoxidation and aziridination of olefins and their synthetic applications. *Chem. Rev.,* **2014**, *114*(16), 8199-8256.
[http://dx.doi.org/10.1021/cr500064w] [PMID: 24785198] h) Sone, T.; Yamaguchi, A.; Matsunaga, S.; Shibasaki, M. Enantioselective synthesis of 2,2-disubstituted terminal epoxides *via* catalytic

asymmetric Corey-Chaykovsky epoxidation of ketones. *Molecules,* **2012**, *17*(2), 1617-1634. [http://dx.doi.org/10.3390/molecules17021617] [PMID: 22314382] i) Iida, T.; Yamamoto, N.; Sasai, H.; Shibasaki, M. New asymmetric reactions using a gallium complex: a highly enantioselective ring opening of epoxides with thiols catalyzed by a gallium,lithium,bis(binaphthoxide) complex. *J. Am. Chem. Soc.,* **1997**, *119*, 4783-4784. [http://dx.doi.org/10.1021/ja9702576]

[17]   a) Palucki M, Pospisil PJ, Zhang W, Jacobsen EN. Highly enantioselective, low-temperature epoxidation of styrene. *J Am Chem Soc.***1994**; *116*: 9333-9334. b) Collman JP, Wang Z, Straumanis A, Quelquejeu M, Rose E. An efficient catalyst for asymmetric epoxidation of terminal olefins. *J Am Chem Soc.***1999**; *121*: 460-461. c) Botes AL, Weijers CAGM, Botes PJ, van Dyk MS. Enantioselectivities of yeast epoxide hydrolases for 1,2-epoxides. *Tetrahedron: Asymmetry,* **1999**;*10*: 3327-3336. d) Goswami A, Totleben MJ, Singh AK, Patel RN. Stereospecific enzymatic hydrolysis of racemic epoxide: a process for making chiral epoxide. *Tetrahedron Asymmetry,* **1999**, *10*, 3167-3175.

[18]   Fiaud, J.C.; Kagan, H.B. Kinetic resolution. In: *Topics in Stereochemistry*; Eliel, E.L.; Wilen, S.H., Eds.; John Wiley and Sons: New York, **2007**; pp. 249-340. Vol. *18*.

[19]   a) Keith, J.M.; Larrow, J.F.; Jacobsen, E.N. Practical considerations in kinetic resolution reactions. *Adv. Synth. Catal.,* **2001**, *343*, 5-26. [http://dx.doi.org/10.1002/1615-4169(20010129)343:1<5::AID-ADSC5>3.0.CO;2-I] b) Robinson, D.E.J.E.; Bull, S.D. Kinetic resolution strategies using non-enzymatic catalysts. *Tetrahedron Asymmetry,* **2005**, *14*, 1407-1446. [http://dx.doi.org/10.1016/S0957-4166(03)00209-X]

[20]   Larrow, J.F.; Schaus, S.E.; Jacobsen, E.N. Kinetic resolution of terminal epoxides *via* highly regioselective and enantioselective ring opening with TMSN$_3$; an efficient, catalytic route to 1,2-amino alcohols. *J. Am. Chem. Soc.,* **1996**, *118*, 7420-7421. [http://dx.doi.org/10.1021/ja961708+]

[21]   Jacobsen, E.N.; Kakuchi, F.; Konsler, R.G.; Larrow, J.F.; Tokunaga, M. Enantioselective catalytic ring opening of epoxides with carboxylic acids. *Tetrahedron Lett.,* **1997**, *38*, 773-776. [http://dx.doi.org/10.1016/S0040-4039(96)02414-8]

[22]   Tokunaga, M.; Larrow, J.F.; Kakiuchi, F.; Jacobsen, E.N. Asymmetric catalysis with water: efficient kinetic resolution of terminal epoxides by means of catalytic hydrolysis. *Science,* **1997**, *277*(5328), 936-938. [http://dx.doi.org/10.1126/science.277.5328.936] [PMID: 9252321]

[23]   Ready, J.M.; Jacobsen, E.N. Asymmetric catalytic synthesis of α-aryloxy alcohols: kinetic resolution of terminal epoxides *via* highly enantioselective ring-opening with phenols. *J. Am. Chem. Soc.,* **1999**, *121*, 6086-6087. [http://dx.doi.org/10.1021/ja9910917]

[24]   a) Larrow, J.F.; Hemberger, K.E.; Jasmin, S.; Kabir, H.; Morel, P. Commercialization of the hydrolytic kinetic resolution of racemic epoxides: toward the economical large-scale production of enantiopure epichlorohydrin. *Tetrahedron Asymmetry,* **2003**, *14*, 3589-3592. [http://dx.doi.org/10.1016/j.tetasy.2003.09.018] b) Liu, Y.; DiMare, M.; Marchese, S.A.; Jacobsen, E.N.; Jasmin, S. Hydrolytic kinetic resolution of epoxides. US Patent 6693206 2004. c) Aouni, L.; Hemberger, K.E.; Jasmin, S.; Kabir, H.; Larrow, J.F.; Le-Fur, I.; Morel, P.; Schlama, T. Industrialization Studies of the Jacobsen Hydrolytic Kinetic Resolution of Epicholohydrin. In: *Asymmetric Catalysis on Industrial Scale: Challenges, Approaches And Solutions*; Blaser, H.U.; Schidmt, E., Eds.; Wiley-VCH Verlag Gmbh & Co: Germany, **2004**; pp. 165-199.

[25]   Schaus, S.E.; Brandes, B.D.; Larrow, J.F.; Tokunaga, M.; Hansen, K.B.; Gould, A.E.; Furrow, M.E.; Jacobsen, E.N. Highly selective hydrolytic kinetic resolution of terminal epoxides catalyzed by chiral (salen)Co(III) complexes. Practical synthesis of enantioenriched terminal epoxides and 1,2-diols. *J. Am. Chem. Soc.,* **2002**, *124*(7), 1307-1315. [http://dx.doi.org/10.1021/ja016737l] [PMID: 11841300]

[26] Nielsen, L.P.C.; Stevenson, C.P.; Blackmond, D.G.; Jacobsen, E.N. Mechanistic investigation leads to a synthetic improvement in the hydrolytic kinetic resolution of terminal epoxides. *J. Am. Chem. Soc.,* **2004**, *126*(5), 1360-1362.
[http://dx.doi.org/10.1021/ja038590z] [PMID: 14759192]

[27] White, D.E.; Tadross, P.M.; Lu, Z.; Jacobsen, E.N. A broadly applicable and practical oligomeric (salen) Co catalyst for enantioselective epoxide ring-opening reactions. *Tetrahedron,* **2014**, *70*(27-28), 4165-4180.
[http://dx.doi.org/10.1016/j.tet.2014.03.043] [PMID: 25045188]

[28] a) Breinbauer, R.; Jacobsen, E.N. Cooperative asymmetric catalysis with dendrimeric [co(salen)] complexes. *Angew. Chem. Int. Ed. Engl.,* **2000**, *39*(20), 3604-3607.
[http://dx.doi.org/10.1002/1521-3773(20001016)39:20<3604::AID-ANIE3604>3.0.CO;2-9] [PMID: 11091412] b) Nielsen, L.P.C.; Zuend, S.J.; Ford, D.D.; Jacobsen, E.N. Mechanistic basis for high reactivity of (salen)Co-OTs in the hydrolytic kinetic resolution of terminal epoxides. *J. Org. Chem.,* **2012**, *77*(5), 2486-2495.
[http://dx.doi.org/10.1021/jo300181f] [PMID: 22292515]

[29] Ready, J.M.; Jacobsen, E.N. Highly active oligomeric (salen)co catalysts for asymmetric epoxide ring-opening reactions. *J. Am. Chem. Soc.,* **2001**, *123*(11), 2687-2688.
[http://dx.doi.org/10.1021/ja005867b] [PMID: 11456948]

[30] Ready, J.M.; Jacobsen, E.N. A practical oligomeric [(salen)Co] catalyst for asymmetric epoxide ring-opening reactions. *Angew. Chem. Int. Ed. Engl.,* **2002**, *41*(8), 1374-1377.
[http://dx.doi.org/10.1002/1521-3773(20020415)41:8<1374::AID-ANIE1374>3.0.CO;2-8] [PMID: 19750769]

[31] White, D.E.; Jacobsen, E.N. New oligomeric catalyst for the hydrolytic kinetic resolution of terminal epoxides under solvent-free conditions. *Tetrahedron Asymmetry,* **2003**, *14*, 3633-3638.
[http://dx.doi.org/10.1016/j.tetasy.2003.09.024]

[32] Belser, T.; Jacobsen, E.N. Cooperative catalysis in the hydrolytic kinetic resolution of epoxides by chiral [(salen)co(iii)] complexes immobilized on gold colloids. *Adv. Synth. Catal.,* **2008**, *350*, 967-971.
[http://dx.doi.org/10.1002/adsc.200800028]

[33] a) Shin, C.K.; Kim, S.J.; Kim, G-J. New chiral cobalt salen complexes containing Lewis acid BF3; a highly reactive and enantioselective catalyst for the hydrolytic kinetic resolution of epoxides. *Tetrahedron Lett.,* **2004**, *45*, 7429-7433.
[http://dx.doi.org/10.1016/j.tetlet.2004.08.067] b) Kim, G.J.; Lee, H.S.; Kim, H.C.; Yun, J.W.; Kim, S.J. Chiral salen catalyst and methods for the preparation of chiral compounds from racemic epoxides by using new catalyst. US Patent 6884750 2005.

[34] Thakur, S.S.; Li, W.; Kim, S.J.; Kim, G-J. Highly reactive and enantioselective kinetic resolution of terminal epoxides with H2O and HCl catalyzed by new chiral (salen)Co complex linked with Al. *Tetrahedron Lett.,* **2005**, *46*, 2263-2266.
[http://dx.doi.org/10.1016/j.tetlet.2005.02.012]

[35] Thakur, S.S.; Li, W.; Shin, C.K.; Kim, G.J. Enantioselective hydrolytic kinetic resolution of 1,2-epoxy-3-phenoxy propane derivatives by new chiral (salen) cobalt complexes. *Catal. Lett.,* **2005**, *104*, 151-156.
[http://dx.doi.org/10.1007/s10562-005-7944-x]

[36] Thakur, S.S.; Li, W.; Shin, C.K.; Kim, G.J. Asymmetric ring opening of terminal epoxides *via* kinetic resolution catalyzed by new bimetallic chiral (salen) co complex. *Chirality,* **2006**, *1*, 37-43.
[http://dx.doi.org/10.1002/chir.20211]

[37] a) Thakur, S.S.; Chen, S.W.; Li, W.; Shin, C.K.; Koo, Y.M.; Kim, G.J. Synthesis of optically pure terminal epoxide and 1, 2-diol *via* hydrolytic kinetic resolution catalyzed by new heterometallic salen complexes. *Synth. Commun.,* **2006**, *36*, 2371-2383.
[http://dx.doi.org/10.1080/00397910600640354] b) Thakur, S.S.; Chen, S.W.; Li, W.; Shin, C.K.;

Koo, Y.M.; Kim, G.J. A new dinuclear chiral salen complexes for asymmetric ring opening and closing reactions: Synthesis of valuable chiral intermediates. *J. Organomet. Chem.,* **2006**, *691*, 1862-1872.
[http://dx.doi.org/10.1016/j.jorganchem.2005.12.044]

[38]   Thakur, S.S.; Li, W.; Kim, S.J.; Kim, G.J. Asymmetric ring opening of some terminal epoxides catalyzed by dimeric type novel chiral Co(salen) complexes. *Stud. Surf. Sci. Catal.,* **2006**, *159*, 205-208.
[http://dx.doi.org/10.1016/S0167-2991(06)81569-7]

[39]   For examples using InCl₃ and TlCl₃.4H₂O as Lewis acid, see (a) Ranu BC. Indium Metal and its halides in organic synthesis. Eur J Org Chem 2000; 13: 2347-2356; (b) Sengupta S, MondalS. InCl3: A new Lewis acid catalyst for reactions with α-diazocarbonyl compounds. Tetrahedron Lett 1999; 40: 8685-8688; (c) Sakae U, Kazuhiro S, Masaya O. Thallium(III) chloride tetrahydrate as a Lewis acid catalyst for aromatic alkylation and acylation. *Bull. Chem. Soc. Jpn.,* **1972**, *45*, 860-863.

[40]   Funabashi, K.; Jachmann, M.; Kanai, M.; Shibasaki, M. Multicenter strategy for the development of catalytic enantioselective nucleophilic alkylation of ketones: Me₂Zn addition to α-ketoesters. *Angew. Chem. Int. Ed. Engl.,* **2003**, *42*(44), 5489-5492.
[http://dx.doi.org/10.1002/anie.200351650] [PMID: 14618585]

[41]   Chen, S.W.; Thakur, S.S.; Li, W.; Shin, C.K.; Koo, Y.M.; Kim, G.J. Efficient catalytic synthesis of optically pure 1, 2- azido alcohols through enantioselective epoxide ring opening with HN₃. *J. Mol. Catal. Chem.,* **2006**, *259*, 116-120.
[http://dx.doi.org/10.1016/j.molcata.2006.06.002]

[42]   Konsler, R.G.; Karl, J.; Jacobsen, E.N. Cooperative asymmetric catalysis with dimeric salen complexes. *J. Am. Chem. Soc.,* **1998**, *120*, 10780-10781.
[http://dx.doi.org/10.1021/ja982683c]

[43]   Patel, D.; Kurrey, G.R.; Shinde, S.S.; Kumar, P.; Kim, G-J.; Thakur, S.S. Dinuclear salen cobalt complex incorporating Y(OTf)₃: enhanced enantioselectivity in the hydrolytic kinetic resolution of epoxides. *RSC Advances,* **2015**, *5*, 82699-82703.
[http://dx.doi.org/10.1039/C5RA12408E]

[44]   Patel, D.; Thakur, S.S.; Shinde, S.S.; Kumar, P. Catalytic and efficient synthesis of valuable optically active terminal epoxides and 1,2-diols using new Lanthanum triflate assisted C1-symmetric bimetallic chiral salen cobalt complex. *Lett. Org. Chem.,* **2019**, *16*, 960-966.

[45]   Kim, GJ; Kim, SJ; Li, W; Chen, SW; Shin, CK; Thakur, SS Synthesis of enantiopure epoxide compounds using dimeric chiral salen catalys *tKorean Chem Eng Res,* **2005**, *43*, 647-661.

[46]   Shin, CK; Ahn, CH; Li, W; Kim, GJ. Catalytic activity of dinuclear chiral salen complexes immobilized on modified SBA-15. *Stud. Surf. Sci. Catal.,* **2007**, 165, 737-740.
[http://dx.doi.org/10.1016/S0167-2991(07)80426-5]

[47]   Venkatasubbaiah, K.; Gill, C.S.; Takatani, T.; Sherrill, C.D.; Jones, C.W. A versatile co(bisalen) unit for homogeneous and heterogeneous cooperative catalysis in the hydrolytic kinetic resolution of epoxides. *Chemistry,* **2009**, *15*(16), 3951-3955.
[http://dx.doi.org/10.1002/chem.200900030] [PMID: 19266511]

[48]   Ford, D.D.; Nielsen, L.P.C.; Zuend, S.J.; Musgrave, C.B.; Jacobsen, E.N. Mechanistic basis for high stereoselectivity and broad substrate scope in the (salen)Co(III)-catalyzed hydrolytic kinetic resolution. *J. Am. Chem. Soc.,* **2013**, *135*(41), 15595-15608.
[http://dx.doi.org/10.1021/ja408027p] [PMID: 24041239]

[49]   a) Kahn, M.G.C.; Weck, M. Highly crosslinked polycyclooctyl-salen Cobalt (III) for the hydrolytic kinetic resolution of terminal epoxides. *Catal. Sci. Technol.,* **2012**, *2*, 386-389.
[http://dx.doi.org/10.1039/C1CY00290B] b) Kahn, M.G.C.; Stenlid, J.H.; Weck, M. Poly(styrene) resin supported Cobalt(III) salen cyclic oligomers as active heterogeneous HKR catalysts. *Adv. Synth. Catal.,* **2012**, *354*, 3016-3024.

[http://dx.doi.org/10.1002/adsc.201200528]

[50]   Love, J.A.; Morgan, J.P.; Trnka, T.M.; Grubbs, R.H. A practical and highly active ruthenium-based catalyst that effects the cross metathesis of acrylonitrile. *Angew. Chem. Int. Ed. Engl.,* **2002**, *41*(21), 4035-4037.
       [http://dx.doi.org/10.1002/1521-3773(20021104)41:21<4035::AID-ANIE4035>3.0.CO;2-I]   [PMID: 12412073]

[51]   Liu, Y.; Rawlston, J.; Swann, A.T.; Takatani, T.; Sherrill, C.D.; Ludovice, P.J.; Weck, M. The bigger, the better: Ring-size effects of macrocyclic oligomeric Co(III)-salen catalysts. *Chem. Sci. (Camb.),* **2011**, *2*, 429-438.
       [http://dx.doi.org/10.1039/C0SC00517G]

[52]   a) Yang, H.; Zhang, L.; Zhong, L.; Yang, Q.; Li, C. Enhanced cooperative activation effect in the hydrolytic kinetic resolution of epoxides on [Co(salen)] catalysts confined in nanocages. *Angew. Chem. Int. Ed. Engl.,* **2007**, *46*(36), 6861-6865.
       [http://dx.doi.org/10.1002/anie.200701747] [PMID: 17668435] b) Yang, H.; Zhang, W.; Yang, Q.; Li, C. Asymmetric ring-opening of epoxides on chiral Co(Salen) catalyst synthesized in SBA-16 through the "ship in a bottle" strategy. *J. Catal.,* **2007**, *248*, 204-212.
       [http://dx.doi.org/10.1016/j.jcat.2007.03.006]

[53]   Rossbach, B.M.; Leopold, K.; Weberskirch, R. Self-assembled nanoreactors as highly active catalysts in the hydrolytic kinetic resolution (HKR) of epoxides in water. *Angew. Chem. Int. Ed. Engl.,* **2006**, *45*(8), 1309-1312.
       [http://dx.doi.org/10.1002/anie.200503291] [PMID: 16425317]

[54]   Oh, C.R.; Choo, D.J.; Shim, W.H.; Lee, D.H.; Roh, E.J.; Lee, S.G.; Song, C.E. Chiral Co(III)(salen)-catalysed hydrolytic kinetic resolution of racemic epoxides in ionic liquids. *Chem. Commun. (Camb.),* **2003**, (9), 1100-1101.
       [PMID: 12772922]

[55]   Kim, D.H.; Shin, U.S.; Song, C.E. Oxidatively pure chiral (salen)Co(III)-X complexes *in situ* prepared by Lewis acid-promoted electron transfer from chiral (salen)Co(II) to oxygen: Their application in the hydrolytic kinetic resolution of terminal epoxides. *J. Mol. Catal. Chem.,* **2007**, *271*, 70-74.
       [http://dx.doi.org/10.1016/j.molcata.2007.02.033]

[56]   Song, C.E. Enantioselective chemo- and bio-catalysis in ionic liquids. *Chem. Commun. (Camb.),* **2004**, (9), 1033-1043.
       [http://dx.doi.org/10.1039/b309027b] [PMID: 15116175]

[57]   Aerts, S.; Weyten, H.; Buekenhoudt, A.; Gevers, L.E.M.; Vankelecom, I.F.J.; Jacobs, P.A. *Chem Commun Recycling of the homogeneous Co-Jacobsen catalyst through solvent-resistent nanofiltration*; SRNF, **2004**, pp. 710-711.

[58]   Aerts, S.; Weyten, H.; Buekenhoudt, A.; Weyten, H.; Vankelecom, I.F.J.; Jacobs, P.A. The influence of solvent choice, acid activation and surfactant addition on the hydrolytic kinetic resolution (HKR) of terminal epoxides. *Tetrahedron Asymmetry,* **2005**, *16*, 657-660.
       [http://dx.doi.org/10.1016/j.tetasy.2004.11.024]

[59]   Berkessel, A.; Ertürk, E. Hydrolytic kinetic resolution of epoxides catalyzed by Chromium(III)*endo,endo* 2,5-diaminonorbornane salen [Cr(III) DIANANE salen] complexes, improved activity,low catalyst loading. *Adv. Synth. Catal.,* **2006**, *348*, 2619-2625.
       [http://dx.doi.org/10.1002/adsc.200606181]

[60]   Beigi, M.; Roller, S.; Haag, R.; Liese, A. Polyglycerol supported Co- and Mn-salen complexes as efficient and recyclable homogeneous catalysts for the hydrolytic kinetic resolution of terminal epoxides and asymmetric olefin epoxidation. *Eur. J. Org. Chem.,* **2008**, 2135-2141.
       [http://dx.doi.org/10.1002/ejoc.200701230]

[61]   Wezenberg, S.J.; Kleij, A.W. Cooperative activation in the hydrolytic kinetic resolution of epoxides by a bis-cobalt(III)salen-calix[4]arene hybrid. *Adv. Synth. Catal.,* **2010**, *352*, 85-91.

[http://dx.doi.org/10.1002/adsc.200900673]

[62]   Park, J.; Lang, K.; Abboud, K.A.; Hong, S. Self-assembly approach toward chiral bimetallic catalysts: bis-urea-functionalized (salen)cobalt complexes for the hydrolytic kinetic resolution of epoxides. *Chemistry,* **2011,** *17*(7), 2236-2245.
[http://dx.doi.org/10.1002/chem.201002600] [PMID: 21294187]

[63]   Zhu, C.; Yuan, G.; Chen, X.; Yang, Z.; Cui, Y. Chiral nanoporous metal-metallosalen frameworks for hydrolytic kinetic resolution of epoxides. *J. Am. Chem. Soc.,* **2012,** *134*(19), 8058-8061.
[http://dx.doi.org/10.1021/ja302340b] [PMID: 22545656]

[64]   Kurahashi, T.; Fujii, H. Unique ligand-radical character of an activated cobalt salen catalyst that is generated by aerobic oxidation of a cobalt(II) salen complex. *Inorg. Chem.,* **2013,** *52*(7), 3908-3919.
[http://dx.doi.org/10.1021/ic302677f] [PMID: 23517550]

[65]   Ren, W-M.; Wang, Y-M.; Zhang, R.; Jiang, J-Y.; Lu, X-B. Mechanistic aspects of metal valence change in SalenCo(III)OAc-catalyzed hydrolytic kinetic resolution of racemic epoxides. *J. Org. Chem.,* **2013,** *78*(10), 4801-4810.
[http://dx.doi.org/10.1021/jo400325f] [PMID: 23631353]

[66]   a) Hong, X.; Billon, L.; Mellah, M.; Schulz, E. Chemical-promoted oxidative polymerization of modified-cobalt salen complexes, efficient catalysts for the dynamic HKR. *Catal. Sci. Technol.,* **2013,** *3*, 723-729.
[http://dx.doi.org/10.1039/C2CY20644G] b) Nimmagadda, S.K.; Mallojjala, S.C.; Woztas, L.; Wheeler, S.E.; Antilla, J.C. Enantioselective synthesis of chiral oxime ethers: desymmetrization and dynamic kinetic resolution of substituted cyclohexanones. *Angew. Chem. Int. Ed. Engl.,* **2017,** *56*(9), 2454-2458.
[http://dx.doi.org/10.1002/anie.201611602] [PMID: 28111889] c) Bhat, V.; Welin, E.R.; Guo, X.; Stoltz, B.M. Bhat V, Welin ER, Guo X, Stoltz BM. Advances in stereoconvergent catalysis from 2005 to 2015: transition-metal-mediated stereoablative reactions, dynamic kinetic resolutions, and dynamic kinetic asymmetric transformations. *Chem. Rev.,* **2017,** *117*(5), 4528-4561.
[http://dx.doi.org/10.1021/acs.chemrev.6b00731] [PMID: 28164696] dHuerta, F.F.; Minidis, A.B.E.; Bäckvall, J-E. Racemisation in asymmetric synthesis. Dynamic kinetic resolution and related processes in enzyme and metal catalysis. *Chem. Soc. Rev.,* **2001,** *30*, 321-331.
[http://dx.doi.org/10.1039/b105464n]

[67]   Matkiewicz, K; Bukowska, A; Bukowski, W Novel highly active polymer supported chiral Co(III)–salen catalysts for hydrolytic kinetic resolution of epichlorohydrin. *J Mol Catal A: Chemical,* **2013,** *368– 369*, 43-52.

[68]   Bredihhina, J.; Villo, P.; Andersons, K.; Toom, L.; Vares, L. Hydrolytic and aminolytic kinetic resolution of terminal bis-epoxides. *J. Org. Chem.,* **2013,** *78*(6), 2379-2385.
[http://dx.doi.org/10.1021/jo3024335] [PMID: 23363444]

[69]   Wu, F.; Wang, K.; Li, Z.; Zhu, X. Design and synthesis of binuclear Co-salen catalysts for the hydrolytic kinetic resolution of epoxides. *Catal. Commun.,* **2015,** *68*, 101-104.
[http://dx.doi.org/10.1016/j.catcom.2015.05.006]

[70]   Kennedy, M.R.; Burns, L.A.; Sherrill, C.D. Counterion and substrate effects on barrier heights of the hydrolytic kinetic resolution of terminal epoxides catalyzed by Co(III)-salen. *J. Phys. Chem. A,* **2015,** *119*(2), 403-409.
[http://dx.doi.org/10.1021/jp511261z] [PMID: 25506779]

[71]   Bukowska, A.; Bukowski, W.; Kleczyńska, S.; Matkiewicz, K. Hydrolytic kinetic resolution of racemic glycidyl esters against a chiral salen complex of cobalt (III) on a polymeric support. *Chemik,* **2016,** *70*, 375-382.

[72]   Dandachi, H.; Zaborova, E.; Kolodziej, E.; David, O.R.P.; Hannedouche, J.; Mellah, M.; Jaber, N.; Schulz, E. Mixing and matching chiral cobalt- and manganese-based calix-salen catalysts for the asymmetric hydrolytic ring opening of epoxides. *Tetrahedron Asymmetry,* **2016,** *27*, 246-253.
[http://dx.doi.org/10.1016/j.tetasy.2016.02.006]

[73]   a) Kureshy, R.I.; Khan, N.H.; Abdi, S.H.R.; Patel, S.T.; Jasra, R.V. Simultaneous production of chirally enriched epoxides and 1,2-diols from racemic epoxides *via* hydrolytic kinetic resolution (HKR). *J. Mol. Catal.*, **2002**, *179*, 73-77.
       [http://dx.doi.org/10.1016/S1381-1169(01)00394-6] b) Kureshy, R.I.; Singh, S.; Khan, N.U.; Abdi, S.H.R.; Ahmad, I.; Bhatt, A.; Jasra, R.V. Improved catalytic activity of homochiral dimeric cobalt-salen complex in hydrolytic kinetic resolution of terminal racemic epoxides. *Chirality,* **2005**, *17*(9), 590-594.
       [http://dx.doi.org/10.1002/chir.20196] [PMID: 16200537]

[74]   Tak, R.; Kumar, M.; Menapara, T.; Gupta, M.; Kureshy, R.I.; Khan, N.H.; Suresh, E. Asymmetric hydrolytic and aminolytic kinetic resolution of racemic epoxides using recyclable macrocyclic chiralCobalt(III) salen complexes. *Adv. Synth. Catal.,* **2017**, *359*, 3990-4001.
       [http://dx.doi.org/10.1002/adsc.201700788]

[75]   Liu, M.; Zhao, Z-P.; Wang, M.; Zhu, P-P. *Proceedings of the Asia-pacific engineering and technology conference (APETC 2017),* **2017**, pp. 671-676.

[76]   Mengozzi, L.; El Garah, M.; Gualandi, A.; Iurlo, M.; Fiorani, A.; Ciesielski, A.; Marcaccio, M.; Paolucci, F.; Samorì, P.; Cozzi, P.G. Phenoxy Al(salophen) scaffolds: Synthesis, electrochemical properties, and self-Assembly at surfaces of multifunctional systems. *Chemistry,* **2018**, *24*(46), 11954-11960.
       [http://dx.doi.org/10.1002/chem.201801118] [PMID: 29603481]

[77]   Zhang, S.; Zhang, B.; Liang, H.; Liu, Y.; Qiao, Y.; Qin, Y. Encapsulation of homogeneous catalysts in mesoporous materials using diffusion-Limited atomic layer deposition. *Angew. Chem. Int. Ed. Engl.,* **2018**, *57*(4), 1091-1095.
       [http://dx.doi.org/10.1002/anie.201712010] [PMID: 29232495]

[78]   Mower, M.P.; Blackmond, D.G. *In-Situ* monitoring of enantiomeric excess during a catalytic kinetic resolution. *ACS Catal.,* **2018**, *8*, 5977-5982.
       [http://dx.doi.org/10.1021/acscatal.8b01411]

[79]   Sun, K.; Li, W-X.; Feng, Z.; Li, C. Cooperative activation in ring-opening hydrolysis of epoxides by Co-salen complexes: A first principle study. *Chem. Phys. Lett.,* **2009**, *470*, 259-263.
       [http://dx.doi.org/10.1016/j.cplett.2009.01.044]

[80]   Kumar, P.; Naidu, V.; Gupta, P. Application of hydrolytic kinetic resolution (HKR) in the synthesis of bioactive compounds. *Tetrahedron,* **2007**, *63*, 2745-2785.
       [http://dx.doi.org/10.1016/j.tet.2006.12.015]

[81]   Hansen, K.B.; Leighton, J.L.; Jacobsen, E.N. On the mechanism of asymmetric nucleophilic ring-opening of epoxides catalyzed by (salen)CrIII complexes. *J. Am. Chem. Soc.,* **1996**, *118*, 10924-10925.
       [http://dx.doi.org/10.1021/ja962600x]

[82]   a) Bartoli, G.; Bosco, M.; Carlone, A.; Locatelli, M.; Melchiorre, P.; Sambri, L. Asymmetric catalytic synthesis of enantiopure N-protected 1,2-amino alcohols. *Org. Lett.,* **2004**, *6*(22), 3973-3975.
       [http://dx.doi.org/10.1021/ol048322l] [PMID: 15496077] b) Bartoli, G.; Bosco, M.; Carlone, A.; Locatelli, M.; Melchiorre, P.; Sambri, L. Direct catalytic synthesis of enantiopure 5-substituted oxazolidinones from racemic terminal epoxides. *Org. Lett.,* **2005**, *7*(10), 1983-1985.
       [http://dx.doi.org/10.1021/ol050675c] [PMID: 15876035] c) Birrell, J.A.; Jacobsen, E.N. A practical method for the synthesis of highly enantioenriched trans-1,2-amino alcohols. *Org. Lett.,* **2013**, *15*(12), 2895-2897.
       [http://dx.doi.org/10.1021/ol401013s] [PMID: 23742206]

[83]   Loy, R.N.; Jacobsen, E.N. Enantioselective intramolecular openings of oxetanes catalyzed by (salen)Co(III) complexes: access to enantioenriched tetrahydrofurans. *J. Am. Chem. Soc.,* **2009**, *131*(8), 2786-2787.
       [http://dx.doi.org/10.1021/ja809176m] [PMID: 19199427]

[84]   Mazet, C.; Jacobsen, E.N. Dinuclear (salen)Al complexes display expanded scope in the conjugate

cyanation of α,β-unsaturated imides. *Angew. Chem. Int. Ed. Engl.,* **2008**, *47*(9), 1762-1765.
[http://dx.doi.org/10.1002/anie.200704461] [PMID: 18219639]

[85]   a) Peretti, K.L.; Ajiro, H.; Cohen, C.T.; Lobkovsky, E.B.; Coates, G.W. A highly active, isospecific cobalt catalyst for propylene oxide polymerization. *J. Am. Chem. Soc.,* **2005**, *127*(33), 11566-11567. [http://dx.doi.org/10.1021/ja053451y] [PMID: 16104709] b) Thomas, R.M.; Widger, P.C.B.; Ahmed, S.M.; Jeske, R.C.; Hirahata, W.; Lobkovsky, E.B.; Coates, G.W. Enantioselective epoxide polymerization using a bimetallic cobalt catalyst. *J. Am. Chem. Soc.,* **2010**, *132*(46), 16520-16525. [http://dx.doi.org/10.1021/ja1058422] [PMID: 21043488]

[86]   Cohen, C.T.; Thomas, C.M.; Peretti, K.L.; Lobkovsky, E.B.; Coates, G.W. Copolymerization of cyclohexene oxide and carbon dioxide using (salen)Co(III) complexes: synthesis and characterization of syndiotactic poly(cyclohexene carbonate). *Dalton Trans.,* **2006**, (1), 237-249. [http://dx.doi.org/10.1039/B513107C] [PMID: 16357982]

CHAPTER 3

Recent Trends in Asymmetric Heterogeneous Flow Catalysis

Rama Jaiswal[1], Melad Shaikh[2] and Kalluri V.S. Ranganath[1,*]

[1] *Department of Chemistry, Institute of Science, Banaras Hindu University, Varanasi, India*

[2] *Indian Institute of Science Education and Research, Tirupati, India*

Abstract: Catalytic reactions play a major role in industry to produce a number of compounds, which are essential in our daily life. These reactions cover synthesis of biofuels from bio waste, oil refining, cracking of hydrocarbons, hydrogenations, dehydrogenations, partial oxidations and fermentations. In surface catalysis, the catalytic reactions occur mainly on the surface, where number of steps involved are adsorption, diffusion and reaction on the surface and desorption of the products. Producing the target product with a high turnover number (TON) and turnover frequency (TOF) is a major challenge for surface catalysis. Recently, flow systems have been developed to produce high quality chemicals and reduce time and energy. In this direction, polymers, metal oxides like alumina and silica, metal nanoparticles of Pt, Pd, and carbon nanotubes (CNTs)have been modified (or treated)with various chiral ligands to synthesize highly active and enantioselective heterogeneous catalysts for the flow process. The aim of this review is to highlight the potential application of flow systems in heterogeneous catalysis. The unique combination of high levels of selectivity in heterogeneous systems together with ease of separation, purification and recyclability makes this heterogeneous system under flow conditions, one of the most promising strategies for the synthesis of fine chemicals on an industrial scale. This review focuses on the most representative examples of this emerging research field, highlights, and future perspectives of flow systems in heterogeneous catalysis. Recent achievements in this area using metal supported and self-supported organo catalysts are discussed.

Keywords: Asymmetric Catalysis, Flow Catalysis, Heterogeneous Catalysis, Supported Catalysts.

INTRODUCTION

Catalysis is a key for the efficient and sustainable synthesis of organic compounds, representing one of the most economical and ecological impacting

---

[*] **Corresponding author Kalluri V.S. Ranganath:** Department of Chemistry, Institute of Science, Banaras Hindu University, Varanasi, India; Tel: 9685458916; E-mails: rangakvs@gmail.com; ranganath.chem@bhu.ac.in

Goutam Kumar Patra & Santosh Singh Thakur (Eds.)

technologies of our recent days [1 - 4]. Limited natural resources and serious environmental issues urgently require catalysts that are more active and highly selective with good stability [5]. The asymmetric heterogeneous catalysis is having many advantages compared to the homogeneous catalysis in terms of recycling and hence in reducing the cost of materials [6, 7]. In homogeneous catalysis, the handling and separation is often troublesome, resulting in undesired (metal) contaminations of the final product. Thus, most reported catalytic systems for asymmetric homogeneous catalysis are not suitable for a commercial application on a large scale.

The asymmetric heterogeneous catalysis plays a major role in obtaining the product with high enantioselectivity. In heterogeneous catalysis, the reaction occurs not only at surfaces but also at interfaces where the reactions occur through adsorption, activation of reactant molecules and desorption of products. Organic as well as inorganic molecules have been utilized to modify the surface of metal or metal oxides to produce highly active materials [8]. Homogeneous catalysts are often quite expensive and neither very robust nor very broad in scope. Heterogeneous catalysis can obviate these disadvantages. Catalytic processes and catalytic reactions can be either homogeneous (use of a soluble catalyst) or heterogeneous (use of an insoluble solid catalyst). On an industrial scale, heterogeneous catalytic processes are highly preferred due to the ease of recovery and reuse of the catalyst, and the simple and clean separation of the catalyst from the reaction products. Thus, supported by those advantages and improved analytical tools, various methods have been developed to immobilize well-known chiral catalysts on solid surfaces. Since the resulting immobilized catalysts react the same way as soluble catalysts, they can be applied to the same reactions with similar or often reduced levels of selectivity and reactivity due to detrimental steric and mass transport effects. This is due to a large proportion of the catalytically active sites being buried deep inside the supporting matrix. Thus, the reactants have limited access to the catalytic sites, which is in stark contrast to easily accessible catalysts in a homogeneous solution. The advantage of a greatly facilitated separation of the immobilized catalysts from the products is often outweighed by the increased cost of the immobilized catalysts compared to the soluble ones. This is especially true when the often-observed low stability of the immobilized catalyst precludes the recycling of the catalyst.

If the immobilized catalysts can sustain the reaction conditions for a long period, they would be of great use for industry. For combinatorial chemistry and high-throughput processes, continuous flow reaction systems have been developed and implemented in many reactions [9 - 13]. In this book chapter, we present and discuss the most representative examples of catalysis under heterogeneous conditions in a fixed bed reactor.

In a continuous flow reactor, the concept of heterogeneous catalysis and a fixed bed system has proven to be an excellent alternative to produce chiral molecules with high yields and ee's [14, 15]. A flow system has several advantages in comparison with a batch procedure such as easy handling, reuse of the catalyst, easy separation, better safety, low cost production and the possibility of operating under harsh conditions in short reaction time (Fig. **1**).

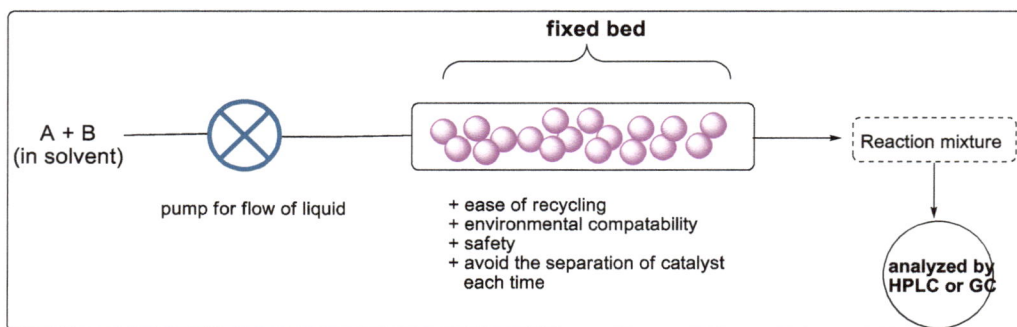

Fig. (1). Schematic diagram of a flow system under heterogeneous conditions.

When considering the removal of the solvent in reactions under flow conditions, it is essential that the reactants and products remain in the liquid phase, under the processing conditions employed; otherwise it will result in the fouling of the catalyst. It is for this reason that continuous flow reactions are more often performed at higher concentrations than their batch counterparts are, but few allow the complete exclusion of solvent. To overcome this issue, additional research is required to improve post reaction separation of solvents in order to further develop solvent recycling for use with continuous flow reactors [16].

Asymmetric Hydrogenation

The enantioselective hydrogenation of keto esters, ketones and olefins have been carried out in continuous flow reactors [17]. In most cases, the catalyst bed (Fig. **2**) consists of a metal (especially either Pt or Pd) adsorbed on a metal oxide surface (in most cases Al_2O_3 and in few cases, SiO_2 or TiO_2). Then, the catalyst surface is modified by adsorption of a chiral inductor, usually an alkaloid.

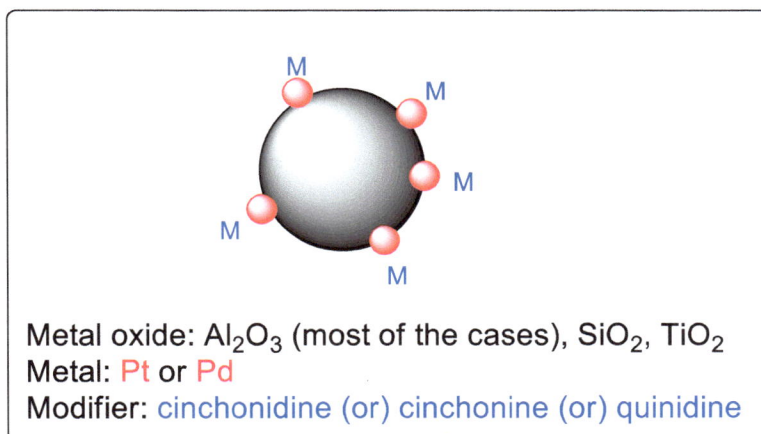

Metal oxide: Al$_2$O$_3$ (most of the cases), SiO$_2$, TiO$_2$
Metal: Pt or Pd
Modifier: cinchonidine (or) cinchonine (or) quinidine

Fig. (2). The heterogeneous catalyst used in a fixed bed reactor in the enantioselective hydrogenation of ketoesters and ketones.

Metal-metal oxide, metal oxide-modifier, metal-modifier interactions play a major role in the enantioselective hydrogenation of ketoesters and ketones. Enantioselective hydrogenations of pyruvates, ketones, and ketopentolactones have been carried out in continuous flow reactors using the catalysts described above (Fig. **2**).

The Pt-catalysed enantioselective hydrogenation of α-keto esters was carried out by admixing a small quantity of a chiral modifier with either silica or alumina and Pt [17d]. The asymmetric hydrogenation of ethyl pyruvate (EP) to ethyl lactate was carried out by mixing a chiral modifier (cinchonidine) to the reactant feed containing EP, provided a good system for a flow process. The fixed bed system consisted of alumina supported Pt. The highest ee achieved in this system was greater than 90% with a flow rate of 4.3 mL/min. Further, a TOF of 84000h^{-1} was achieved, which corresponds to the production rate of 1.94 mol/min/g of Pt. Interestingly, a fixed bed reactor exhibited better activity (4.3 mol/min/L) than a batch reactor (0.06-0.33 mol/min/L) in the hydrogenation of EP [18].

Li and co-workers carried out the enantioselective hydrogenation of ethyl-2-oxo-4-phenylbutyrate on a cinchonidine modified Pt/γ-Al$_2$O$_3$ catalysts in a fixed bed reactor [19]. The chiral product, (*R*)-(+)-ethyl-2-hydroxy-4-phenylbutyrate, was obtained in a 95% conversion and 68% ee. The cinchonidine ligand (Scheme **1**) was adsorbed on the surface (pre-prepared chiral catalyst) prior to its use in the fixed bed reactor. In this method, the competitive adsorption of solvent and reactant molecules plays an important role, as ee values obtained in toluene (up to

68%) were higher than those obtained in other solvents such as acetic acid and ethanol (22 to 65%), which are more strongly adsorbed and, in a given solvent, the ee decreases over time. By comparison, when a toluene solution of both the chiral modifier and EP were flown through the reactor containing Pt/γ-Al$_2$O$_3$ (*in-situ* adsorption of the cinchonidine ligand on the Pt/γ-Al$_2$O$_3$ surface), the highest ee obtained was 45% [19]. As the chiral modifier was flowing along with the reactant, it took a long time for the modifier to be adsorbed on the surface and generate enough active catalyst. This might be the reason for the decrease of ee of the product.

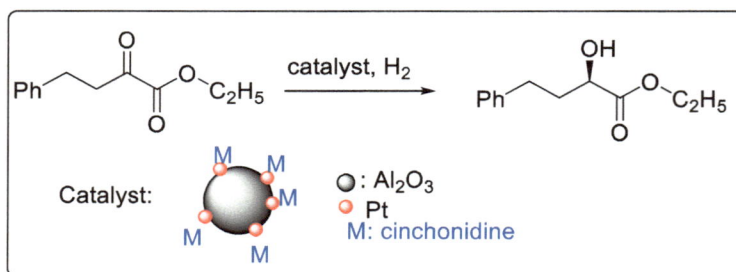

Scheme 1. Enantioselective hydrogenation of ethyl-2-oxo-4-phenylbutyrate on cinchonidine-modified Pt/γ-Al$_2$O$_3$ [19].

Enantioselective hydrogenation of isopropyl-4,4,4-trifluoroacetoacetate to the corresponding alcohol was also carried out by Baiker and co-workers [20] in a fixed bed reactor over Pt/Al$_2$O$_3$. The chiral modifier was *O*-methyl-cinchonidine. A solution of the reactant, additive and chiral modifier in THF was flown through the fixed bed reactor containing a mixture of Pt/Al$_2$O$_3$ and Al$_2$O$_3$ (Scheme 2). The addition of a small amount of trifluoro acetic acid caused a two-fold increase of the ee but a significant decrease in the rate of hydrogenation. This decrease in rate was attributed to the formation of a ligand H$^+$-EP complex on the surface, which was less mobile on the surface. This might result in a decrease in the rate of hydrogenation because of the lower probability of encounter between the chiral complex and adsorbed hydrogen. Higher enantioselectivitie's (60-90%, *R*-enantiomer) could be achieved at temperatures lower than 0°C and the highest ee was obtained at -10 °C with a TOF of 340h^{-1} (Scheme 2) [20]. The asymmetric hydrogenation of ethyl benzoylformate to ethyl mandalate in ethanol in a fixed bed reactor was reported at 0 and 50 °C using the highly active Pt/Al$_2$O$_3$ catalyst pre-modified with cinchonidine. A 52% ee (*R* enantiomer) was obtained at 50 °C [21]. An unexpected inversion was observed in this reaction after the removal of soluble fraction of the modifier at 50°C. Reversal of enantioselectivity (-18% ee, *S*-enantiomer) was observed after washing the catalyst with ethanol, which was attributed to the alteration of Pt crystallites and to a support effect [21]. Pt-alumina interactions play a major role in establishing the enantioselectivity in the hydrogenation of EP. Morphological changes of Pt/Al$_2$O$_3$ did occur upon thermal treatment and were also due to alkaloid-metal-alumina interactions [21, 22].

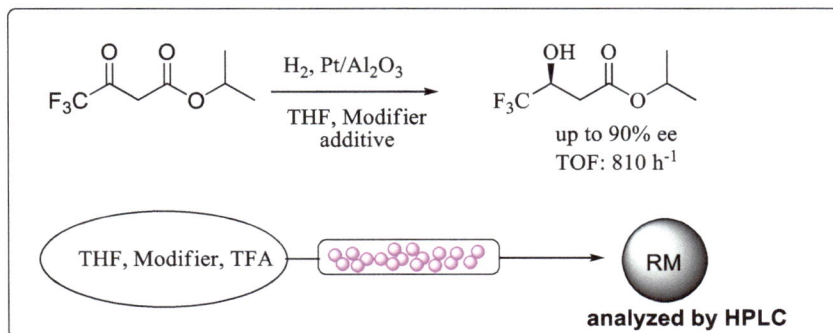

Scheme 2. Enantioselective hydrogenation of isopropyl-4,4,4-trifluoroacetoacetate on Pt/γ-Al$_2$O$_3$ using *O*-methyl-cinchonidine as an modifier [20].

A chiral pre-modified catalyst was used effectively in the enantioselective hydrogenation of EP. The asymmetric hydrogenation of EP, carried out in a continuous flow reactor at 25°C, on a 1% Pt-cinchonidine/SiO$_2$ catalyst, has been reported by Ruiz and co-workers [23]. A moderate enantiomeric excess (63%) was observed at a 10 bar hydrogen pressure. A mixture EP (0.01 M) and cinchonidine (CD) (0.034 M) in hexane was allowed to pass though the fixed bed at a flow rate of 0.3 mL/min. A comparative study on the ee in continuous and batch reactors was reported for the hydrogenation of EP. The maximum ee observed in a fixed bed reactor was 89.9%, which is slightly lower than the 92% ee obtained in the batch reactor. The enantioselective hydrogenation of 2,2,2-trifluoroacetophenone over Pt/Al$_2$O$_3$ modified by cinchona alkaloids was carried out in a mixture of toluene/AcOH in the presence of an additive, trifluoroacetic acid. The hydrogenation reactions were carried out in an H-cube high-pressure continuous flow system at a flow rate of 1.0 mL/min with a trifluoroacetic concentration of 1.0 mM. Inversion of selectivity was observed with Pt-cinchonine and Pt-Quinodine catalysts in the absence of the additive. However, these results were not observed in the presence of the additive [24].

The asymmetric hydrogenation of methyl-2-acetamidoacrylate (MAA) was reported using a rhodium complex supported on phosphotungstic acid (PTA)/Al$_2$O$_3$. The heterogeneous rhodium complex, [Rh(COD)(*S*)-MonoPhos)$_2$]/PTA/Al$_2$O$_3$, was used in the asymmetric hydrogenation of MAA in a continuous flow reactor under 1 bar hydrogen pressure. The maximum enantioselectivity observed in the hydrogenation of MAA was 97% with 99% conversion after 12 h. The hydrogenated compound(*R*)-*N*-acetylalanine was produced at a rate of 0.174g/h giving a total of 2.1 g after 12 h. The Rh-complex was immobilized on neutral gamma Al_2O_3 and also on mesoporous Al_2O_3. A higher activity and selectivity were observed with [Rh(COD)(*S*)-MonoPhos)$_2$]/PTA/Al$_2$O$_3$ (mesoporous) than with [Rh(COD)(*S*)-MonoPhos)]/PTA/Al$_2$O$_3$ (neutral gamma) due to a larger surface area and pore volume when compared with the former. Similar metal complexes were used to generate metal organic frame work materials, which were used as self-supported catalysts [25a]. These self-supported catalysts were used in both batch reactor and continuous flow system as a stationary phase for the asymmetric hydrogenation of α-dehydroamino acid methyl esters [25b]. The hydrogenation reaction was run continuously for 144 h with > 99% conversion providing 97% enantioselectivity. This self-supporting strategy allowed simple and efficient catalyst immobilization without the use of additional support such as Al_2O_3 or silica.

Asymmetric Arylation

Enantioselective arylation of aldehydes with organozinc reagents appears as the most convenient methodology to generate secondary alcohols [26]. After the pioneering work of Seebach co-worker [27] on the asymmetric catalytic phenyl addition to aldehydes using a highly reactive agent, PhTi(O^{i}Pr)$_3$, chemists have shown a continuous interest to develop an asymmetric version. Fu and co-workers [28] reported the first asymmetric arylation reaction, the phenylation of *p*-chorobenzadehyde (99% yield, 57% ee), using a chiral azaferrocene as catalyst in a batch procedure. Recently, the enantioselective arylation of aldehydes with aryl zinc reagents using heterogeneous catalyst was reported in a flow reactor system (Scheme **3**) [29]. Chiral amino alcohols derived from proline and pyrrolidine supported on a Merrifield resin were employed as heterogeneous catalysts in the arylation of aldehydes using phenyl ethyl zinc in a fixed bed reactor. Both these reactants were dissolved in toluene and were passed through a flow reactor. The residence time and temperature affected significantly the rate and enantioselectivity of the arylation reaction. Stability of the catalyst in the flow system is greater than in the batch system. The former showed no significant loss of activity even after eight cycles, whereas, with the latter, significant loss of activity was observed after the third cycle [29]. Various aldehydes shown in Scheme **3** were tested under the flow reactor conditions. The best ee's were obtained with benzaldehydes bearing an electron-donating substituent at the para position.

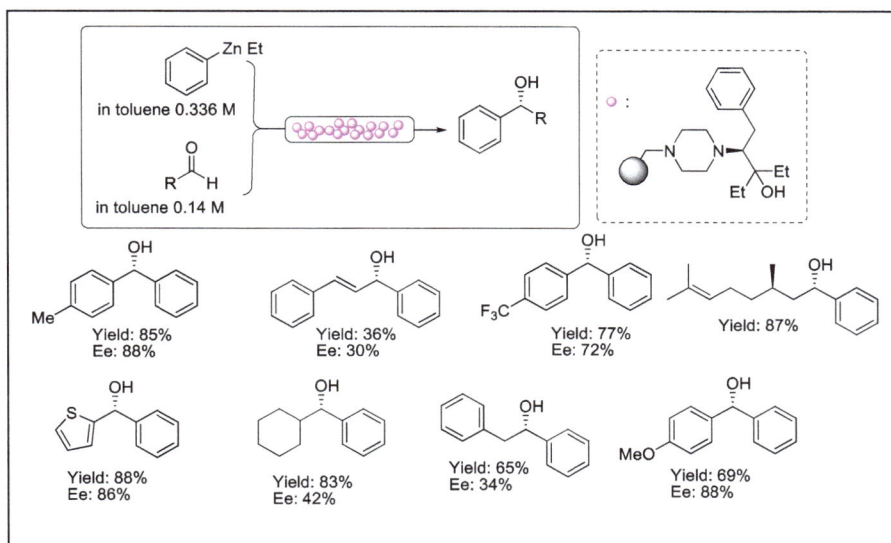

Scheme 3. Asymmetric arylation of aldehydes with diethyl zinc catalyzed by proline and pyrrolidine derivatives supported on a Merrifield resin [29].

Alkylation

The asymmetric addition of diethyl zinc to an aldehyde, using a catalyst consisting of ephedrine supported on a chloromethylated polystyrene-divinylbenzenes (*meta and para*) copolymer, was reported by Hodge and co-workers [30]. At the optimum flow rate, the hydrogenation of benzaldehyde in toluene gave chiral 1-phenylpropan-1-ol in high yields (96-98%) and enantioselectivitie's (97-98%) in a bench-top flow reactor. This heterogeneous catalyst was prepared by using grafting methods. When the homogeneous catalyst, (1*R*,2*S*)-*N*-benzylephedrin, was used in a batch system with toluene as solvent, phenylpropanol was obtained with a 80% ee and 88% yield, average values [30]. The higher ee in the flow system is attributed to the higher concentration of the reactants and the continuous removal of the product, which may interact with diethyl zinc to promote racemisation. Using polystyrene-polydivinyl supported camphor derivatives for the same reaction afforded phenylpropanol in a high yield (81%) and ee (94%) for the first 108h. After 275 h, the yield dropped to 60% and the ee to 81%. Luis and co-workers [31], for the same reaction at room temperature using polymeric monoliths containing chiral amino alcohols, reported a higher ee (99%) and quantitative conversion (85% selectivity to phenyl propanol) of benzaldehyde. The mixture was re-circulated through the catalyst column for 24 h. Polymeric heterogeneous catalyst prepared from a mixture of monomers (10% chiral alcohol and 90% divinylbenzene) in toluene/1-dodecanol was used in monolithic column and provided a better enantioselectivity (99%) than that obtained with homogeneous analogues (90%). This may be due to the formation of more appropriate asymmetric cavities during the preparation of the catalyst *via* polymerization.

Pericàs and co-workers [32] also reported the addition of diethyl zinc to benzaldehydes in toluene, in a continuous flow system using, as catalyst, a cross-linked polystyrene to which the amino alcohol (*R*)-2-(1-piperazinyl)-1-1,2-triphenylethanol has been grafted. With a series of benzaldehydes, a high conversion (98-100%), a high selectivity (98-100%), and a high ee (89-93%) were obtained.

Asymmetric Aldol Reactions

The asymmetric aldol and Michael reactions are one of the most important C–C bond formation reactions, since the resultant compounds are used in the synthesis of drug molecules and natural products. Various homogeneous and heterogeneous organo catalysts give good yields and ee's in the asymmetric aldol reaction [33, 34]. Odedra and Seeberger [35] reported the 5-(pyrrolidin-2-yl) tetrazole-catalyzed aldol condensation of acetone with aromatic aldehydes, in DMSO and at 50 °C, in a flow micro reactor. The yields of aldol varied from 36 to 78% and the ee's (S enantiomer) from 57 to 75%. The aldol reaction between cyclohexanone and 4-cynobenzaldehyde gave a mixture of diastereoisomers (*syn/anti* = 48:52) in 86% yield and ee's of 77% (*syn*) and 81% (*anti*). A tripeptide attached to a resin (H-Pro-Pro-Asp-resin) was used as an efficient catalyst for the reaction of aromatic aldehydes with acetone [36]. The flow synthesis was carried out in DMSO, at room temperature, and at a pressure of 60 bar. The best conversion and yield were obtained with benzaldehydes bearing a strong electron-withdrawing substituent (100% conversion, 99% yield) with ee's of 74-80%. With the other aromatic aldehydes investigated, the yields were lower due to the dehydration of the aldol (27-71%). In the case of 2-fluorobenzaldehyde, the dehydration of aldol was predominant, which is due to the fact that "F" is strongly electron withdrawing by induction (quantitative conversion) but electron donating by resonance. The later effect favors the dehydration of the aldol. The ee's of the β-hydroxyketones were in the 70-80% range. The catalyst showed a very good stability. For the reaction of acetone with 4-nitrobenzaldehyde, after twenty cycles of reuse of the catalyst (80 h), there was no change of the yield and of the ee. A TON of 710 was observed overall in the continuous flow process, which is, much better than the TON calculated from the batch process (472) [36]. Wennemers and co-workers [37] investigated catalysts made of the same tripeptide attached to various resins in the reaction of acetone with aromatic and aliphatic aldehydes. The best catalyst was the tripeptide hooked to a polyethylene glycol–polyacrylamide resin. The yields and ee's of the aldol were respectively 72% and 72% with benzaldehyde, 94% and 80% with 4-nitrobenzaldehyde, 70-76% and 72% with RCHO (R = propyl, *i*-propyl, cyclohexyl). Even the hindered *neo*-pentyl aldehyde afforded the corresponding aldol in 42% yield and 70% ee. Reversal of enantioselectivity of the aldol from (*R*) (24%) to (*S*) (42%) was observed as the peptide length increases from a di- to a tripeptide for the reaction of acetone with 2-nitrobenzaldehyde (78-87% conversion) [38]. In the case of *iso*-propanal, the reversal was more pronounced: 80% (*R*) vs 67% (*S*) but the conversion was very low (18-13%). The catalytic enantioselective reaction of a benzaldehyde with cyclohexanone under flow conditions using as heterogeneous catalyst, a 5-(pyrrolidin-2-yl) tetrazole functionalized silica is shown in Scheme **4**, **I**. Cavazzini, Massi and co-workers [39] carried out this reaction in a 0.22 M solution, in toluene, of a benzaldehyde bearing an electron withdrawing group and cyclohexanone (0.66 M). Performing the reaction at 50 °C and at a flow rate of 5 μL/min in a packed bed micro reactor, afforded a mixture of *anti* and *syn* chiral diastereoisomers of the aldol (d.r's of 3:1 and 2:1). The conversion was excellent ($\geq$ 95%) with ee's of 68-82% for the *anti*-isomer. The productivity was estimated at 0.236-0.256 mmol h^{-1} mmolcatalyst^{-1}. In a preliminary report [40], the same group studied the effect of temperature for the asymmetric aldol condensation between cyclohexanone and 4-nitrobenzaldehyde (3:1 molar ratio) in toluene, using the silica supported proline-derived organo catalyst **IV** (Scheme **4**) at a flow rate of 5 μL/min. From 0 to 50 °C, the conversion increased from 38 to 82%. There was a little effect of the temperature on the stereoselectivity: dr *anti/syn* of 5:1 to 4:1, ee$_{anti}$ of 82% to 78%. The productivity increased from 0.27 at 0 °C to 0.59 at 50 °C. A further increase of temperature to 70 °C resulted in the loss of activity, the conversion dropping from 95% to 60% after 90 min. The ee$_{anti}$ was not affected (78%). The silica supported catalyst **IV** showed better activity and enantioselectivity than catalyst **III** (Scheme **4**) at 50 °C: 50% conversion and ee$_{anti}$ of 40% (same dr). A subsequent paper by the same group [41] reported a reaction-progress kinetic analysis coupled with thermodynamic data from nonlinear chromatographic measurements for the asymmetric aldol condensation between cyclohexanone and 4-nitrobenzaldehyde in toluene, at 25 °C, in the same micro reactor, using the silica supported proline-derived organo catalyst **IV** (Scheme **4**). The results reveal two features influencing a heterogeneous continuous-flow process the dependence of the reaction order on the feed composition and on the saturation capacity of the catalytic bed. Pericàs and co-workers [42] have studied polystyrene immobilized proline derivatives as catalysts for the asymmetric aldol reactions in continuous flow reactions. This reaction was carried out in a mixture of solvents (DMF/H$_2$O) between benzaldehydes and cyclohexanone. Catalyst **II** (Scheme **4**) was the best catalyst. Benzaldehydes bearing an electron-withdrawing substituent gave the best yield of the aldol: 85-96% compared to 28-41% for benzaldehydes with an electron-donating substituent. The *anti/syn* diastereoselectivity was high in all cases (90:10 to 98:2) as well as the ee (97 to > 99%). The reaction with 4-nitrobenzaldehyde performed under batch conditions gave similar diastereoselectivity (*anti/syn* ratio of 95:5) and enantioselectivity (96-99%) as compared to the flow conditions. The conversion for the batch experiments dropped from 92% to 60% after seven runs of 24 h. In the batch process, 10 mol% of catalyst was required whereas, in the flow process, it was reduced to 1.6%. The TON observed in the continuous process was 61 compared to 9 in the batch process The authors assumed that the polystyrene supporting catalyst **II** would be more resistant to collapse under continuous-flow conditions than under batch conditions because of the higher level of cross-linking of the polymer, which contains 8% of divinylbenzene.

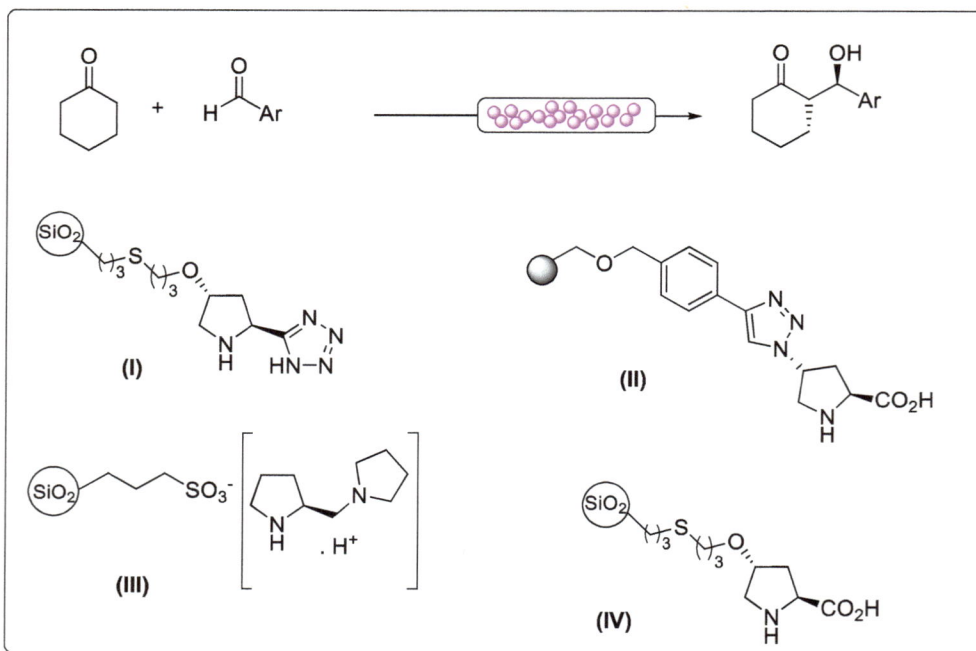

Scheme 4. Asymmetric aldol reaction catalyzed by proline-derived catalysts under heterogeneous flow conditions.

Sels and co-workers [43] reported the asymmetric aldol reaction of 2-butanone with 4-(trifluoromethyl) benzaldehyde. The heterogeneous catalyst used is chiral diamine, was derived from an amino acid, immobilized on an acid support (sulfonated fluoropolymer (Nafion® SR50) in a fixed bed system. The reaction was performed by passing a preheated (45 °C) solution of reagents (aldehyde in 2-butanone 0.125M) through the column at a flow rate of 1.0 mL/h. After 4 h, the yield of aldol was 37% with an ee of 97% for the major *syn* diastereoisomer. After 11h, the yield reached 77% and the ee was 94%. A variety of acid additives were tested in a batch system, at room temperature, and Nafion® SR50 was found to be the best additive affording the aldol in a 87% yield, with a dr (*syn/anti*) of 5:2, and an ee$_{syn}$ of 94%. Electrostatic interactions between the protonated primary amine groups, the conjugate base of the acid are quite effective in this process. Both organic and inorganic sulfonated solid supports were used for the development of acid-base interactions. In this process, the solid acid has a dual function acting not only as anchor for immobilization but also governing the activity and selectivity of the catalyst. Various inorganic supports like zeolites, silica, amberlyst, nafion and organic sulfonated acid such as trifluoroacetic acid (TFA) were used.

Asymmetric Michael Reactions

A variety and versatile heterogeneous catalysts were developed for the asymmetric Michael addition of aldehyde enols/enolates to nitro olefins. Among them, solid supported peptidic organo catalysts were tested first under batch conditions [44] and then applied to flow reactors [45]. Chiral γ-nitroaldehydes were synthesized in excellent yields and ee's under the optimized conditions shown in Scheme **5**. This catalyst (H- Pro-Pro-Asp-NH-resin) was the best of those investigated. The resin was a polystyrene containing 4-methylbenzhydrylamine. The linear aldehydes gave the best yields (60-91%). Benzenepropanal gave a good yield (65%) and very high dr (*syn/anti* of 36:1), whereas a low yield of 22% was observed with 3-methylbutanal (R = *i*Pr). In general, yields and stereoselectivities are also very good or even better in the batch reactions carried out at room temperature and atmospheric pressure [45]. However, the batch reactions require 12 to 24 h whereas, in the flow system (0.1 mL/min), a residence time of only 7.0 min is required to achieve nearly the same results. The fine tuning of the flow rate influenced not only the yield, but also the diastereoselectivity; the lower the flow rate, the lowest the diastereomeric ratio. Authors suggested this is due to the catalyst could affect the epimerization of the product, and consequently, the longer the residence time, the lower the diastereoselectivity.

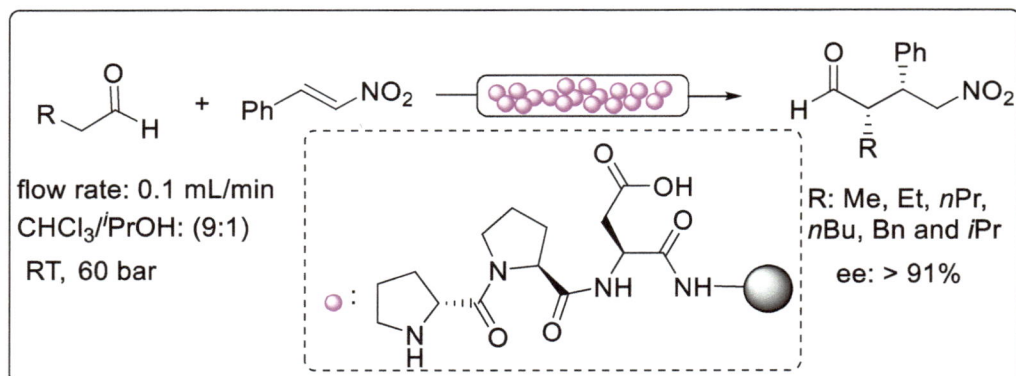

Scheme 5. Asymmetric Michael addition reaction catalyzed by polymer supported peptide catalyst [45]. [*The ee's are those of the syn diastereomer: The dr's (dr: syn/anti) are very good: 11:1 to 22: 1*].

A highly recyclable and stable chiral catalyst was used for the Michael addition of nitro methane to unsaturated ketones under flow catalysis [46]. This chiral catalyst was the polystyrene-supported 9-amino (9-deoxy)*epi*-cinchona alkaloid shown in Scheme **6** showing the Michael addition of ethyl nitroacetate to 4-phenyl-buten-2-one. The conversion of the enone was 98%, the diastereoisomeric ratio 1.1:1, and the ee 96%/97%, which is really remarkable in a heterogeneous system [46]. Both the reactants along with benzoic acid as additive were flowed through the reactor at a flow rate of 50μL/min and at 30 °C. The residence time was 40 min and the TON 36. There was no deterioration of the catalyst after 21 h. The flow process was applied to a large variety of Michael donors and of enones with very good results: conversion of 80% to > 90% and ee's higher than 90% in most cases. The high enantioselectivity may be due to the formation of non-covalent acid-base interactions (electrostatic interaction between the protonated primary amine of the catalyst and a benzoate anion) [47].

In a similar manner, these peptide type or proline type heterogeneous catalysts have been successfully used in the asymmetric Mannich, α-amination and α-aminoxylation reactions [48 - 51]. As a result, various α-functionalized aldehydes and ketones were prepared using flow process. A polymer supported cinchonidine catalyst was first time used in the 1,4-addition reaction by Hodge and co-workers [52] in bench top flow system. A keto ester and methyl vinyl ketone was pumped individually to the fixed tube containing a polymer supported catalyst. The chiral product was obtained in 51% ee in good yields at a flow rate of 7.0 mL/h, having a residence time of 6 h. This result compares well with the results achieved in batch conditions with the same polymer supported cinchonidine (47% ee) and with cinchonidine itself (53% ee). However, authors did not report the TON in both cases (Scheme 7).

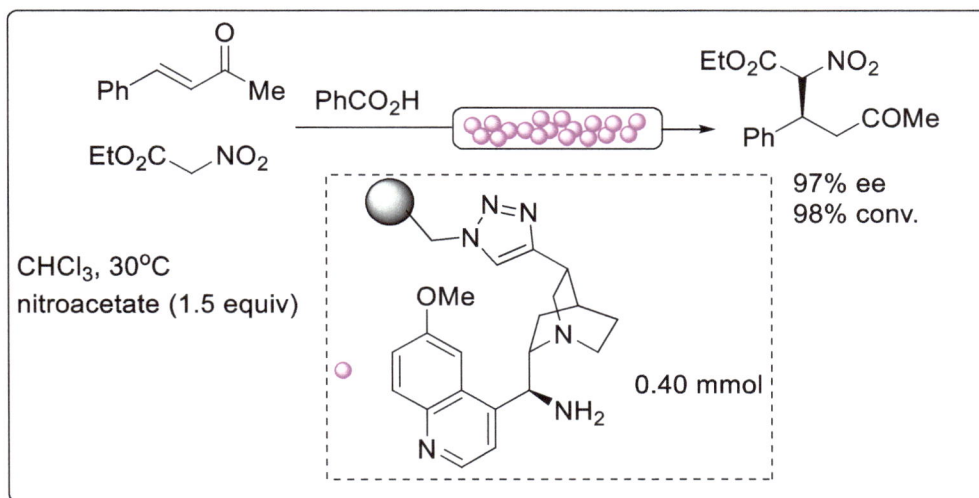

Scheme 6. Asymmetric Michael addition reaction catalyzed by an alkaloid-resin catalyst [46].

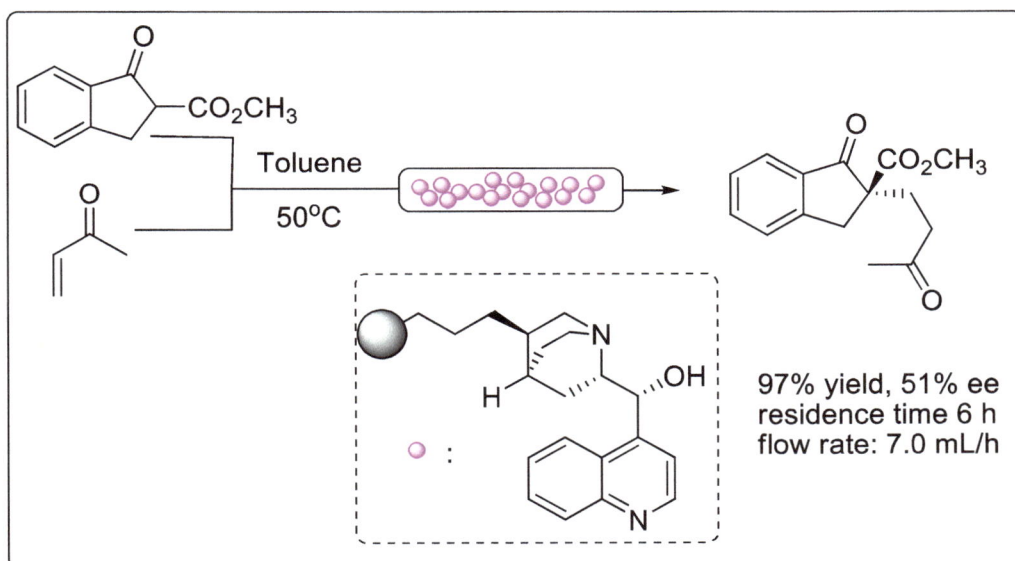

Scheme 7. Enantioselective 1,4-addition reaction catalysed by polymer supported alkaloid catalyst [52].

Alkaline earth metals are very rare in asymmetric catalysis using heterogeneous conditions even in the batch process. In this direction few reports are available based on chiral Mg and Ca complexes in asymmetric reactions [53, 54]. The versatile and robust heterogeneous chiral Ca catalysts were developed by

Kobayashi and co-workers [55]. They developed chiral catalyst, through the reaction of Pybox-CaCl$_2$, which immobilized on polystyrene. Thus, obtained robust and highly active catalyst was tested in a continuous flow system in the asymmetric 1,4-addition of 1,3-dicarbonyl compound with nitro alkenes. The TON observed was 228 with >90% ee.

The enantioselective synthesis of protected (*S*)-mandelonitrile from benzaldehyde has been carried out in a flow reactor. The hydrophobic surface of vinyl benzene beads was modified to hydrophilic surface by allowing free vinyl groups to react with 4-(hydroxybutyl)vinyl ether in a free radical-mediated reaction. The titanium (IV) complex of a polymer supported chiral salen ligand was used as a catalyst in the continuous flow reactor [56]. The asymmetric cyanation of benzaldehyde with TMSCN was carried out under flow system (flow rate: 0.8µL/min). The product, TMS protected mandelonitrile was obtained with 92% conversion and 72% ee was observed. This result is comparable under batch conditions (91% conversion with 73% ee) after 3h of the reaction at room temperature [57]. However, when acetyl cyanide was used, the yield of acetylated (*S*)-mandelonitrile was lower (48% yield, 70% ee).

The immobilization of the chiral catalyst would allow not only recycling, cost reducing but would also be useful as a catalyst bed in a continuous flow reactor as catalyst can reside in the reactor for long period. The reduction of imines with trichlorosilane was reported at room temperature using polystyrene supported picolonamide catalyst. The residence time plays a significant effect towards enantioselectivity (range from 85 to 73%), however, yields were nearly same (from 92 to 94% ee) [58]. Further, 1-(*m*-benzyloxyphenyl)-ethylamine, which is the precursor of a drug used in neuropathic pain was synthesized using the flow reactor and it was isolated in 82% yield with 83% ee in first 3 h. The yields and ee's retained for another 2h and then dropped slowly.

The confinement of carbon nanotubes (CNTs) inside the channel is one of the recent trends in asymmetric heterogeneous catalysis [59, 60]. The asymmetric *anti-selective* nitroaldol reaction in a flow system was developed by Shibasaki and co-workers using self-assembled solid phase catalyst in the confined space of CNTs [61]. The solid phase catalysts prepared by mixing of NdCl$_3$.6H$_2$O and amide based ligands were used in a fixed bed. The stainless steel column filled with the catalyst was immersed in a cryogenic reactor, which could be operated at -40°C and substrates were passed through the flow system by two syringe pumps at a flow rate of 0.6 mL/h. The nitro aldol product was observed with 95% ee (*anti*), 81% yield and TON was 1661 (Scheme **8**).

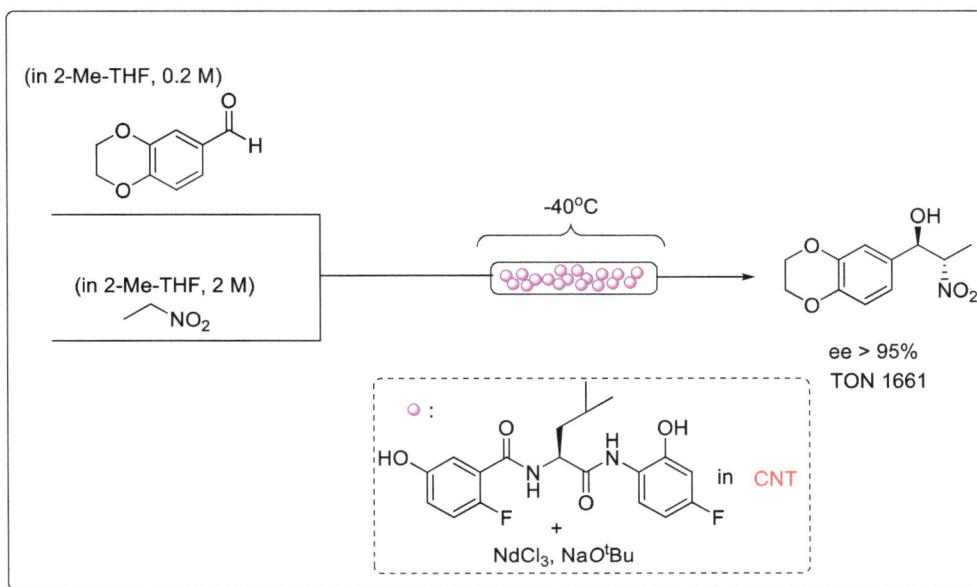

Scheme 8. *anti*-Selective asymmetric nitroaldol reaction in a flow reactor [61].

N-heterocyclic carbenes (NHC) is a well-known versatile and organo catalyst for many reactions such as umpouling, Michael and aldol reactions. Its heterogenization has a lot of advantages such as recycling of the expensive catalyst to reduce the cost [62]. The versatile and privileged chiral triazolium carbene catalyst developed by Rovis [63, 64] which was immobilized on polystyrene showed the high level of enantioselectivity in a flow reactor system at a flow rate of 10µL/min. The intermolecular stetter reaction was carried out to get the optically active chromanones continuously in excellent yields (95%) and enantioselectivitie's (81-95% ee). TON observed in this case was 132 and moreover flow system proved better than the batch system as earlier system produced 2-fold increase of chromanones [65]. In the same reaction, polystyrene supported chiral triazolium catalyst gave good results compared to silica supported catalysts (Scheme **9**).

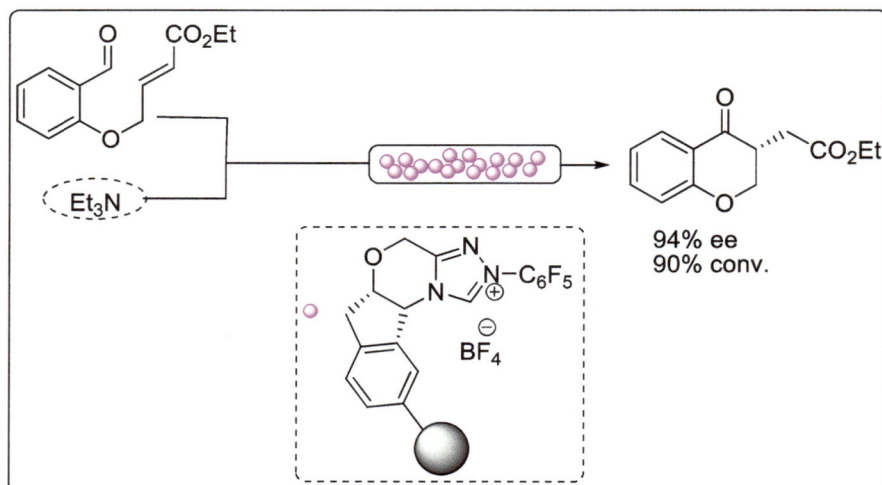

Scheme 9. Asymmetric Stetter reaction catalyzed by polystyrene supported Rovis catalyst (triazolium) in a continuous flow reactor [65].

Introducing a polysilane-supported heterogeneous palladium/alumina hybrid catalyst, Kobayashi and co-workers [66] reported the development of a fixed bed reactor for the hydrogenation of unsaturated C–C bonds and protection of carbobenzyloxy functional groups under solvent-free conditions. Using this method, the authors were able to perform selective hydrogenations without the need of dilute substrates, affording high TON (8700). In the case of solid substrates, small quantities of solvent were required, however, substrate concentrations in the range of 0.33 to 1.0 M were typically employed. The hydrogenation of nitro group proceeded quantitatively under neat conditions at 70 °C with a flow rate of 0.1 mL/min with a TON of 2100.

The alcohol oxidation is one of the most important organic transformations since the resultant carbonyl compounds have been used in various applications. Traditional methods involve the use of toxic, expensive and stoichiometric oxidants and harmful organic solvents [67]. To avoid the use of expensive catalysts, toxic reagents, recyclable heterogeneous catalysts have been developed for the oxidation of alcohols [68]. The Au NPs supported on SiO_2@Yneused for oxidation of secondary alcohols in flow reactor [69]. The oxidation of secondary alcohol was carried out at 90°C using H_2O_2 as an oxidant in an aqeous medium. The flow reactor packed with active gold catalyst and fed the reaction mixture with a flow rate of 0.1 mL/min. The heterogeneous catalyst was able to work continuously for 50 h under flow conditions (Scheme **10**). A comparative study was also done in the oxidation of alcohol using Au NPs on SiO_2@Yne, Al_2O_3@Yne and TiO_2@Yne. Although all the three catalysts were highly active, Au/SiO_2@Yne shows the better activity in the oxidation of alcohols.

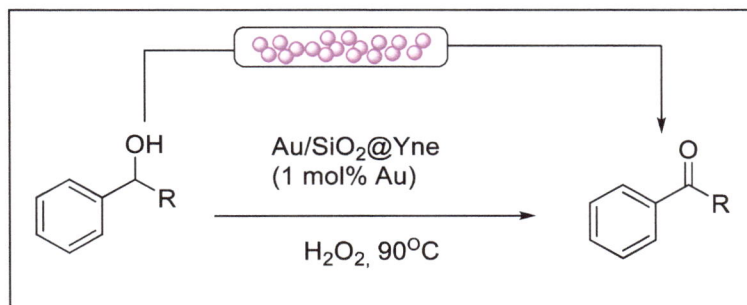

Scheme 10. Oxidation of secondary alcohols catalyzed by Gold supported catalyst in a continuous flow reactor [69].

The stereo selective organo catalyzed Diels–Alder reactions in the fixed bed reactor containing silica supported imidazolidinone were reported by Benaglia and co-workers [70]. This catalyst was packed into an empty stainless steel HPLC column. It efficiently promotes stereo selective Diels–Alder reactions on different substrates for more than 150 h of continuous operation. Regeneration of the column was accomplished and the life of the catalyst was prolonged for 300 h. Three types of commercially available silica (8, 10 and 25μm) were effectively used as supports and however, it was found that 8 and 10 μm silica were good supports

Asymmetric Carbene Transfer Reactions

Heterogeneous copper-catalyzed carbene transfer reactions under flow conditions have been well studied. Carbene transfer reaction of ethyl diazoacetate with olefins using copper catalyst was studied. In this reaction, heterogeneous catalyst, polystyrene supported triazoloyl methyl derivatives were used. Asymmetric cyclopropanation reaction of styrene with ethyl diazoacetate under continuous flow system was developed by Luis and co-workers [71]. The highly active and robust catalysts, polystyrene-supported Pybox-Ru complexes were used. To know the effect of solvent, this reaction was carried was out in dichloromethane, in supercritical CO_2 and also in the absence of solvent. TONs observed in the cyclopropanation reaction in dichloromethane, solvent free and supercritical CO_2 were 6, 35 and 46, respectively. In super critical CO_2, 91% conversion was observed with 65% yield and enantioselectivity (77% ee *trans* and 50% ee *cis* respectively). Later, Luis and co-workers [72] also reported the same reaction using monolithic reactor impregnated Polystyrene-supported Cu-Box under continuous flow systems (Scheme **11**). In the presence of supercritical CO_2, significant increase in product was observed in cyclopropanation reactions.

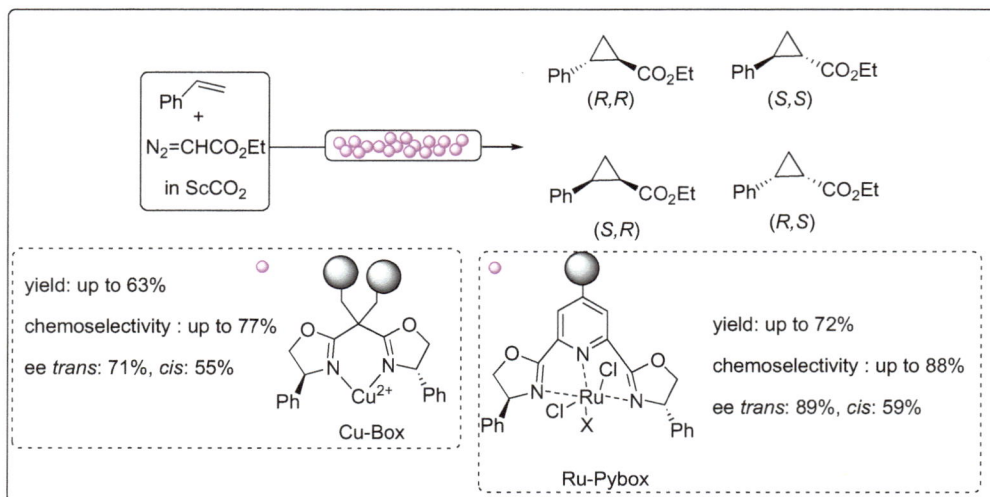

Scheme 11. Asymmetric carbene transfer reaction catalyzed by Cu and Ru-Box ligands [72].

Other Asymmetric Reactions in a Continuous Flow

The self-supported chiral titanium cluster containing binol ligands in a two-step ligand exchange and hydrolysis sequence was used in the Strecker-type reaction. This heterogeneous catalyst was mixed with a celite powder and packed into the column reactor. The Strecker-type reaction of *N*-benzhydrylimine with TMSCN was carried out in a fixed bed reactor and achieved 80% conversion with 96% ee [73]. This heterogeneous catalyst was quite active for at least 85 h (Scheme **12**). Further, three component reaction has also been carried out using this self-organized catalyst in a flow system. Excess cyanide is required to get the maximum yields and ee's (96%) and later it was quenched by saturated solution of ferrous sulphate.

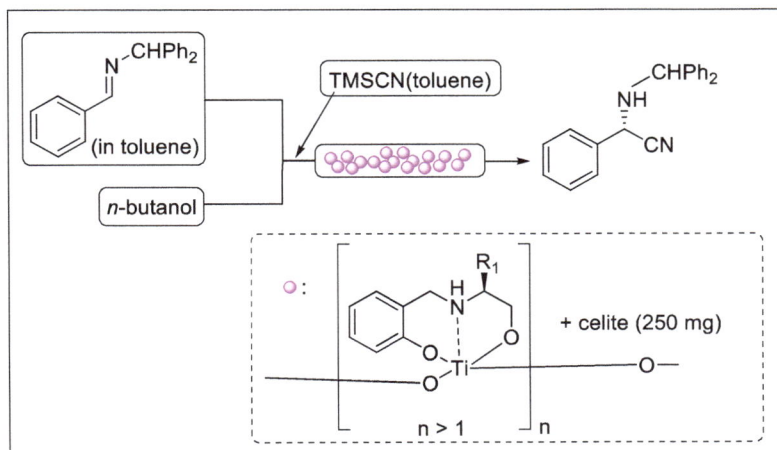

Scheme 12. Self supported titanium catalyst for the Strecker type reaction [73].

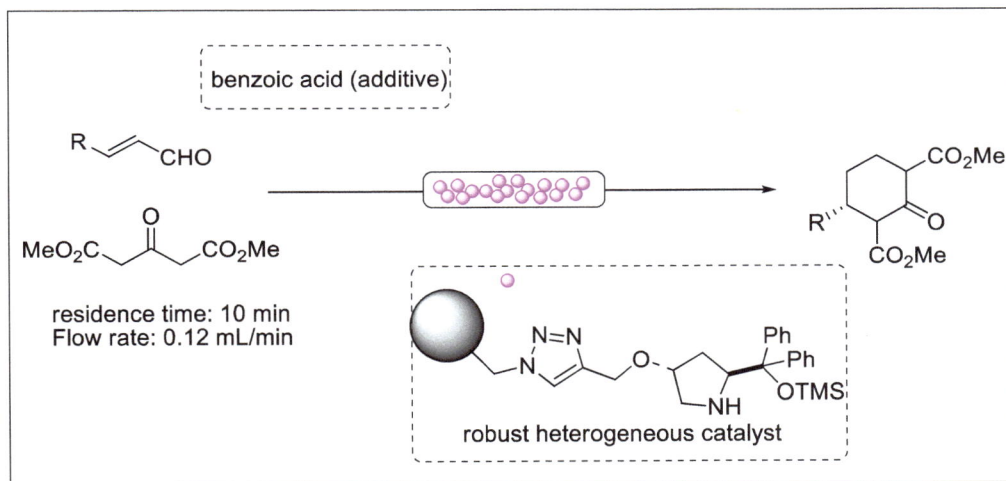

Scheme 13. Polystyrene supported prolinol catalyst in the continuous flow reactor for Robinson annulations [74].

Robinson annulations, consisting of Michael addition and subsequent intramolecular aldol condensation are one of the most well-known reactions. Recently, Pericás and co-workers reported a Robinson annulations reaction in a fixed bed reactor containing diphenylprolinol immobilized on polystyrene catalyst [74]. The reaction between dimethyl-3-oxoglutarate and 4-methoxycinnamaldehyde was carried out to produce functionalized cyclohexanone under heterogeneous conditions (Scheme **13**). This highly robust heterogeneous catalyst was not deteriorated for three days to provide the compound with 97% ee. The use of continuous flow process achieved a TON of 66, which was ten times greater than the batch process.

CONCLUSION AND PERSPECTIVES

In this book chapter, we highlighted the recent examples of asymmetric catalysis under flow conditions [75]. Various materials like alumina, silica, polystyrene, MOFs and recently CNTs have been used as support in heterogeneous asymmetric catalysis. The large interfacial areas generated under continuous reactions are major advantages compared to batch systems. Various catalytic reactions are becoming available in a continuous flow reactor, it is still necessary to develop more asymmetric reactions with high ee's. These reactions in flow reactors generally favour improved yields and enantioselectivity. For example, asymmetric hydrogenations of ketones and ketoesters have been well studied using alkaloids as modifiers on the surface of metal supported on metal oxide. However, hydrogenation of heteroaromatics like pyridine and their derivatives were reported by a few authors under continuous flow reactors. Asymmetric C−C bond formation, oxidations, Friedal-Crafts alkylation reactions were also reported. Reactions in flow catalysis offer safer, more economical and space saving method than conventional batch approaches. It represents a great advantage of flow catalysis with respect to heat transfer, mass transfer issues to overcome the reactivity problems. In addition, flow catalysis can avoid the contamination of catalyst from the reaction mixture, which can re-use without requiring purification process. Continuous monitoring of the reaction progress using different techniques like GC and HPLC are major tasks. Online analytical reaction techniques may be able to perform better in analysing the reaction mixture.

CONSENT FOR PUBLICATION

Not applicable.

CONFLICT OF INTEREST

The author declares that there is no conflict of interest in this chapter.

ACKNOWLEDGEMENTS

This work was partially supported by the DST-SERB New Delhi (project grant file number CRG/2019/000105).

REFERENCES

[1]   Heitbaum, M.; Glorius, F.; Escher, I. Asymmetric heterogeneous catalysis. *Angew. Chem. Int. Ed. Engl.,* **2006**, *45*(29), 4732-4762.
[http://dx.doi.org/10.1002/anie.200504212] [PMID: 16802397]

[2]   Jacobsen, E.N.; Pfaltz, A.; Yamamoto, H. *Comprehensive Asymmetric Catalysis I-III*; Springer: Berlin, **1999**.
[http://dx.doi.org/10.1007/978-3-642-58571-5]

[3]   Sheldon, R.A. E factors, green chemistry and catalysis: an odyssey. *Chem. Commun. (Camb.),* **2008**, (29), 3352-3365.
[http://dx.doi.org/10.1039/b803584a] [PMID: 18633490]

[4]   Sheldon, R.A. Why green chemistry and sustainability of resources are essential to our future. *J. Environ. Monit.,* **2008**, *10*(4), 406-407.
[http://dx.doi.org/10.1039/b801651h] [PMID: 18385859]

[5]   Trindade, A.F.; Gois, P.M.P.; Afonso, C.A.M. Recyclable stereoselective catalysts. *Chem. Rev.,* **2009**, *109*(2), 418-514.
[http://dx.doi.org/10.1021/cr800200t] [PMID: 19209946]

[6]   Schrader, I.; Neumann, S.; Sulce, A.; Schmidt, F.; Azov, V.; Kunz, S. Asymmetric heterogeneous catalysis: Transfer of molecular principles to nanoparticles by ligand functionalization. *ACS Catal.,* **2017**, *7*, 3979-3987.
[http://dx.doi.org/10.1021/acscatal.7b00422]

[7]   Ranganath, K.V.S.; Shaikh, M.; Sahu, A.; Sarvani, G. Catalytic activity of functionalized spinels. *Curr. Org. Chem.,* **2017**, *21*, 2573-2584.
[http://dx.doi.org/10.2174/1385272821666170517144231]

[8]   Neouze, M-A.; Schubert, U. Surface modification of metal and metal oxide nanoparticles by organic ligands. *Monatsh. Chem.,* **2008**, *139*, 183-195.
[http://dx.doi.org/10.1007/s00706-007-0775-2]

[9]   Kapteijn, F.; Moulijn, J.A. *Handbook of Heterogeneous Catalysis,* 2nd ed; Wiley-VCH: Weinheim, **2008**, pp. 2019-2045.

[10]   Tsubogo, T.; Ishiwata, T.; Kobayashi, S. Asymmetric carbon-carbon bond formation under continuous-flow conditions with chiral heterogeneous catalysts. *Angew. Chem. Int. Ed. Engl.,* **2013**, *52*(26), 6590-6604.
[http://dx.doi.org/10.1002/anie.201210066] [PMID: 23720303]

[11]   Masuda, K.; Ichitsuka, T.; Koumura, N.; Sato, K.; Kobayashi, S. Flow fine synthesis with heterogeneous catalysts. *Tetrahedron,* **2018**, *74*, 1705-1730.
[http://dx.doi.org/10.1016/j.tet.2018.02.006]

[12]   Atodiresei, I.; Vila, C.; Rueping, M. Asymmetric organocatalysis in contineous flow: Opportunities for imparting industrial catalysis. *ACS Catal.,* **2015**, *5*, 1972-1985.
[http://dx.doi.org/10.1021/acscatal.5b00002]

[13]   Porta, R.; Benaglia, M.; Puglisi, A. Flow chemistry: Recent developments in the synthesis of pharmaceutical products. *Org. Process Res. Dev.,* **2016**, *20*, 2-25.
[http://dx.doi.org/10.1021/acs.oprd.5b00325]

[14]   Tsubogo, T.; Oyamada, H.; Kobayashi, S. Multistep continuous-flow synthesis of (*R*)- and (*S*)-rolipram using heterogeneous catalysts. *Nature,* **2015**, *520*(7547), 329-332.
[http://dx.doi.org/10.1038/nature14343] [PMID: 25877201]

[15]   a) Gross, E.; Liu, H-C. J.; Toste, F. D.; Somorjai, G. A. Control of selectivity in heterogeneous catalysis by tuning nanoparticle properties and reactor residence time. *Nature. Chem.,* **2012**, *4*: 947-952. b) Seki, T.; Grunwadt, J.-D.; Baiker, A. Heterogeneous catalytic hydrogenations in supercritical fluids: Potential and limitations. *Ind. Eng. Chem. Res.,* **2008**, *47*, 4561-4585.

[16]   Zhao, Y.; Gao, F.; Chen, L.; Garland, M. Chiral fixed bed reactor for stereoselective heterogeneous catalysis: Modification, regeneration, and multiple product syntheses. *J. Catal.,* **2004**, *221*, 274-287.
[http://dx.doi.org/10.1016/S0021-9517(03)00153-2]

[17]   a) Abdallah, R.; Fumey, B.; Meille, V.; de Bellefon, C. Micro-structured reactors as a tool for chiral modifier screening in gas–liquid–solid asymmetric hydrogenations. *Catal. Today.*, **2007**, *125*: 34-39. b) Szöllösi, G.; Cserenyi, S.; Fülöp, F.; Bartók, M. New data to the origin of rate enhancement on the Pt-cinchona catalyzed enantioselective hydrogenation of activated ketones using continuous-flow fixed-bed reactor system. *J. Catal.*, **2008**, *260*: 245-253. c) Cserenyi, S.; Szöllösi, G.; Szori, K.; Fülöp, F.; Bartók, M. Reversal of the ee in enantioselective hydrogenation of activated ketones in continuous-flow fixed-bed reactor system. *Catal. Commun.*, **2010**, *12*: 14-19. d) Baiker, A. Progress in asymmetric heterogeneous catalysis: Design of novel chirally modified platinum metal catalysts. *J. Mol. Catal. A.*, **1997**, *115*, 473-493.

[18]   Kunzle, N.; Soler, J-W.; Baiker, A. Continuous enantioselective hydrogenation in fixed-bed reactor: towards process intensification. *Catal. Today,* **2003**, *79-80*, 503-509.
[http://dx.doi.org/10.1016/S0920-5861(03)00075-0]

[19]   Li, X.; Li, C. Enantioselective hydrogenation of ethyl-2-oxo-4-phenylbutyrate on cinchonidine-modified Pt/γ-Al₂O₃ catalyst using a fixed bed reactor. *Catal. Lett.*, **2001**, *77*, 251-254.
[http://dx.doi.org/10.1023/A:1013280116781]

[20]   Kunzle, N.; Mallat, T.; Baiker, A. Enantioselective hydrogenation of isopropyl-4,4,--trifluoroacetoacetate in a continuous flow reactor. *Appl. Catal. A.*, **2003**, *238*, 251-257.
[http://dx.doi.org/10.1016/S0926-860X(02)00349-6]

[21]   Gao, F.; Chen, Li.; Garland, M. A new origin for stereo-differentiation in the Orito reaction: Residual chiral induction and enantiomeric excess reversal during piecewise continuous experiments using a chiral fixed bed reactor. *J. Catal.,* **2006**, *238*, 402-411.
[http://dx.doi.org/10.1016/j.jcat.2005.12.018]

[22]   Szollosi, G.; Cserenyi, S.; Fülöp, F.; Bartok, M. New data to the origin of rate enhancement on the Pt-cinchona catalyzed enantioselective hydrogenation of activated ketones using continuous-flow fixed bed reactor system. *J. Catal.,* **2008**, *260*, 245-253.
[http://dx.doi.org/10.1016/j.jcat.2008.10.004]

[23]   Ruiz, D.R.; Reyes, P. Enantioselective hydrogenation of ethyl pyruvate in flow reactor over Pt-CD/SiO₂ catalysts. *J. Chil. Chem. Soc.,* **2008**, *53*, 1740-1742.
[http://dx.doi.org/10.4067/S0717-97072008000400024]

[24]   Szöllösi, G.; Cserenyi, S.; Bucsi, I.; Bartok, T.; Fülöp, F.; Bartók, M. Origin of the rate enhancement and enantio differentiation in the heterogeneous enantioselective hydrogenation of 2,2,2-trifluoroacetophenone over Pt/alumina studied in continuous-flow fixed-bed reactor system. *Appl. Catal.,* **2010**, *382*, 263-271.
[http://dx.doi.org/10.1016/j.apcata.2010.05.003]

[25]   a) Madarasz, J. A continuous flow system for asymmetric hydrogenation using supported chiral catalysts. *J. Flow. Chem.*, **2011**, *2*, 62-67. b) Shi, L.; Wang, X.; Sandoval, C.A.; Wang, Z.; Li, H.; Wu, J.; Yu, L.; Ding, K. Development of a continuous-flow system for asymmetric hydrogenation using self-supported chiral catalysts. *Chemistry,* **2009**, *15*, 9855-9867.

[26]   Ramón, D.J.; Yus, M. Chiral tertiary alcohols made by catalytic enantioselective addition of unreactive zinc reagents to poorly electrophilic ketones? *Angew. Chem. Int. Ed. Engl.,* **2004**, *43*(3), 284-287.
[http://dx.doi.org/10.1002/anie.200301696] [PMID: 14705080]

[27]   Weber, B.; Seebach, D. Ti-TADDOLate-catalyzed, highly enantioselective addition of alkyl- and aryl-titanum derivatives to aldehydes. *Tetrahedron,* **1994**, *50*, 7473-7484.
[http://dx.doi.org/10.1016/S0040-4020(01)90475-2]

[28]   Dosa, P.I.; Ruble, J.C.; Fu, G.C. Planar–Chiral heterocycles as ligands in metal-catalyzed processes: Enantioselective addition of organozinc Reagents to aldehydes. *J. Org. Chem.,* **1997**, *62*(3), 444-445.
[http://dx.doi.org/10.1021/jo962156g] [PMID: 11671428]

[29]   Forni, J.A.; Novaes, L.F.T.; Galaverna, R.; Pastre, J.C. Novel polystyrene-immobilized chiral amino

alcohols as heterogeneous ligands for the enantioselective arylation of aldehydes in batch and continuous flow regime. *Catal. Today,* **2018**, *308*, 86-93.
[http://dx.doi.org/10.1016/j.cattod.2017.08.055]

[30]   Hodge, P.; Sung, D.W.L.; Stratford, P.W. Asymmetric synthesis of 1-phenylpropanol using polymer supported chiral catalyst in simple bench top flow systems. J. Chem. Soc. *Perkin Trans.,* **1999**, *1*, 2335-2342.
[http://dx.doi.org/10.1039/a902967b]

[31]   Burguete, M.I. García-Verdugo, E.; Vicent, M. J.; Luis, S. V.; Pennemann, H.; Keyserling, N. G.; Murtens, J. New supported β-amino alcohols as efficient catalysts for the enantioselective addition of diethyl zinc to benzaldehyde under flow conditions. *Org. Lett.,* **2002**, *4*, 3947-3950.
[http://dx.doi.org/10.1021/ol026805o] [PMID: 12599499]

[32]   Pericàs, M.A. Herreĺĺas, C. I.; Sora, L. Fast and enantioselective production of 1-aryl-1-propanols through a single pass, continuous flow process. *Adv. Synth. Catal.,* **2008**, *350*, 927-932.
[http://dx.doi.org/10.1002/adsc.200800108]

[33]   Hartikka, A.; Arvidsson, P.I. 5-(Pyrrolidin-2-yl) tetrazole: Rationale for the increased reactivity of the tetrazole analogue of proline in organocatalyzed aldol reactions. *Eur. J. Org. Chem.,* **2005**, 4287-4295.
[http://dx.doi.org/10.1002/ejoc.200500470]

[34]   Berkessel, A.; Gröger, H. *Asymmetric Organocatalysis: From Biomimetic Concepts to Applications in Asymmetric Synthesis*; Wiley-VCH: Weinheim, **2005**, pp. 130-244.
[http://dx.doi.org/10.1002/3527604677.ch6]

[35]   Odedra, A.; Seeberger, P.H. 5-(Pyrrolidin-2-yl)tetrazole-catalyzed aldol and mannich reactions: acceleration and lower catalyst loading in a continuous-flow reactor. *Angew. Chem. Int. Ed. Engl.,* **2009**, *48*(15), 2699-2702.
[http://dx.doi.org/10.1002/anie.200804407] [PMID: 19288478]

[36]   Ötvös, S.B.; Màndity, I.M.; Fülöp, F. Asymmetric aldol reaction in a continuous flow reactor catalyzed by a highly re-usable heterogeneous peptide. *J. Catal.,* **2012**, *295*, 179-185.
[http://dx.doi.org/10.1016/j.jcat.2012.08.006]

[37]   Revell, J.D.; Gantenbein, D.; Krattiger, P.; Wennemers, H. Solid-supported and pegylated H-Pro-P-o-Asp-NHR as catalysts for asymmetric aldol reactions. *Biopolymers,* **2006**, *84*(1), 105-113.
[http://dx.doi.org/10.1002/bip.20393] [PMID: 16245260]

[38]   Gurka, A.; Bucsi, I.; Kovacs, L.; Szollosi, G.; Bartok, M. Reversal of the enantioselectivity in Aldol addition over immobilized di and tripeptides: Studies under continuous flow conditions. *RSC Advances,* **2014**, *4*, 61611-61618.
[http://dx.doi.org/10.1039/C4RA07188C]

[39]   Bortoloni, O.; Caciolli, L.; Cavazzini, A.; Costa, V.; Greco, R.; Massi, A.; Pasti, L. Silica-supported 5-(pyrrolidin-2-yl) tetrazole: Development of organocatalytic processes from batch to continuous-flow conditions. *Green Chem.,* **2012**, *14*, 992-1000.
[http://dx.doi.org/10.1039/c2gc16673a]

[40]   Massi, A.; Cavazzini, A.; Del Zoppo, L.; Pandoli, O.; Costa, V.; Pasti, L.; Giovannini, P.P. Toward the optimization of continuous-flow aldol and α-amination reactions by means of proline-functionalized silicon packed-bed micro reactors. *Tett. Lett.,* **2011**, *52*, 619-622.
[http://dx.doi.org/10.1016/j.tetlet.2010.11.157]

[41]   Bortolini, O.; Cavazzini, A.; Giovannini, P.P.; Greco, R.; Marchetti, N.; Massi, A.; Pasti, L. A combined kinetic and thermodynamic approach for the interpretation of continuous-flow heterogeneous catalytic processes. *Chemistry,* **2013**, *19*(24), 7802-7808.
[http://dx.doi.org/10.1002/chem.201300181] [PMID: 23589216]

[42]   Ayats, C.; Henseler, A.H.; Pericàs, M.A. A solid-supported organocatalyst for continuous-flow enantioselective aldol reactions. *ChemSusChem,* **2012**, *5*(2), 320-325.
[http://dx.doi.org/10.1002/cssc.201100570] [PMID: 22442839]

[43]   Demuynck, A.L.W.; Peng, L.; de Clippel, F.; Vanderleyden, J.; Jacobs, P.A.; Sels, B.F. Solid acids as heterogeneous support for primary amino acid☐derived diamines in direct asymmetric aldol Reactions. *Adv. Synth. Catal.,* **2011**, *353*, 725-732.
[http://dx.doi.org/10.1002/adsc.201000871]

[44]   Weisner, M.; Revell, J.D.; Wennemers, H. Tripeptides as efficient asymmetric catalysts for 1,4-addition reactions of aldehydes to nitroolefins-A rational approach. *Angew. Chem. Int. Ed.,* **2008**, *47*, 1871-1874.
[http://dx.doi.org/10.1002/anie.200704972]

[45]   Otvös, S.B.; Mándity, I.M.; Fülöp, F. Highly efficient 1,4-addition of aldehydes to nitroolefins: organocatalysis in continuous flow by solid-supported peptidic catalysts. *ChemSusChem,* **2012**, *5*(2), 266-269.
[http://dx.doi.org/10.1002/cssc.201100332] [PMID: 22298413]

[46]   Izquierdo, J.; Ayats, C.; Henseler, A.H.; Pericàs, M.A. A polystyrene-supported 9-amino(9-deoxy)*epi* quinine derivative for continuous flow asymmetric Michael reactions. *Org. Biomol. Chem.,* **2015**, *13*(14), 4204-4209.
[http://dx.doi.org/10.1039/C5OB00325C] [PMID: 25723553]

[47]   Luo, S.; Li, J.; Zhang, L.; Xu, H.; Cheng, J-P. Noncovalently supported heterogeneous chiral amine catalysts for asymmetric direct aldol and Michael addition reactions. *Chemistry,* **2008**, *14*(4), 1273-1281.
[http://dx.doi.org/10.1002/chem.200701129] [PMID: 18000996]

[48]   Ötvös, S.B.; Szloszár, A.; Mándity, I.M. Heterogeneous di peptide catalyzed α-amination of aldehydes in a continuous-flow reactor: Effect of residence time on enantioselectivity. *Adv. Synth. Catal.,* **2015**, *357*, 3671-3680.
[http://dx.doi.org/10.1002/adsc.201500375]

[49]   Fan, X.; Sayalero, S.; Pericàs, M.A. Asymmetric α-amination of aldehydes catalyzed by PS-diphenylprolinol silyl ethers: Remediation of catalyst deactivation for continuous flow operation. *Adv. Synth. Catal.,* **2012**, *354*, 2971-2976.
[http://dx.doi.org/10.1002/adsc.201200887]

[50]   Cambeiro, X.C.; Martín-Rapún, R.; Miranda, P.O.; Sayalero, S.; Alza, E.; Llanes, P.; Pericàs, M.A. Continuous-flow enantioselective α-aminoxylation of aldehydes catalyzed by a polystyrene-immobilized hydroxyproline. *Beilstein J. Org. Chem.,* **2011**, *7*, 1486-1493.
[http://dx.doi.org/10.3762/bjoc.7.172] [PMID: 22238521]

[51]   Martin-Rapün, R.; Sayalero, S.; Pericàs, M.A. Asymmetric anti-Mannich reactions in continuous flow. *Green Chem.,* **2013**, *15*, 3295-3301.
[http://dx.doi.org/10.1039/c3gc41444b]

[52]   Bonfils, F.; Cazaux, I.; Hodge, P.; Caze, C. Michael reactions carried out using a bench-top flow system. *Org. Biomol. Chem.,* **2006**, *4*(3), 493-497.
[http://dx.doi.org/10.1039/B515241K] [PMID: 16446807]

[53]   Choudary, B.M.; Ranganath, K.V.S.; Pal, U.; Kantam, M.L.; Sreedhar, B. Nanocrystalline MgO for asymmetric Henry and Michael reactions. *J. Am. Chem. Soc.,* **2005**, *127*(38), 13167-13171.
[http://dx.doi.org/10.1021/ja0440248] [PMID: 16173743]

[54]   Kumaraswamy, G.; Sastry, M.N.V.; Jena, V.; Kumar, K.R.; Viramani, M. Enantioenriched (*S*)-6,6--diphenylBinol-Ca: A novel and efficient chirally modified metal complex for asymmetric epoxidation of α,β-unsaturated enones. *Tetrahedron Asymmetry,* **2003**, *14*, 3797-3803.
[http://dx.doi.org/10.1016/j.tetasy.2003.08.022]

[55]   Tsubogo, T.; Yamashita, Y.; Kobayashi, S. Toward efficient asymmetric carbon-carbon bond formation: continuous flow with chiral heterogeneous catalysts. *Chemistry,* **2012**, *18*(43), 13624-13628.
[http://dx.doi.org/10.1002/chem.201202896] [PMID: 22968991]

[56]   Belokon, Y.N.; Caveda-Cepas, S.B.; Ikonnikov, S.; Khrustalev, V.N.; Larichev, V.S.; Moscalenko, M.A.; North, M.; Orizu, C.; Tararov, V.I.; Tasinazzo, M.; Timofeeva, G.I.; Yashkina, L.V. The asymmetric addition of trimethylsilyl cyanide to aldehydes catalyzed by chiral (salen) titanium complexes. *J. Am. Chem. Soc.,* **1999**, *121*, 3968-3973.
[http://dx.doi.org/10.1021/ja984197v]

[57]   Lundgren, S.; Ihre, H.; Moberg, C. Enantioselective cyanation of benzaldehyde: An asymmetric polymeric catalyst in a micro reactor. *ARKIVOC,* **2008**, 73-80.

[58]   Porta, R.; Benaglia, M.; Annunziata, R.; Puglisi, A.; Celentano, G. Solid supported chiral *N*-Picolylimidazolidinones: Recyclable catalysts for the enantioselective, metal free and hydrogen free reduction of imines in batch and in flow mode. *Adv. Synth. Catal.,* **2017**, *359*, 2375-2382.
[http://dx.doi.org/10.1002/adsc.201700376]

[59]   Chen, Z.; Guan, Z.; Li, M.; Yang, Q.; Li, C. Enhancement of the performance of a platinum nanocatalyst confined within carbon nanotubes for asymmetric hydrogenation. *Angew. Chem. Int. Ed. Engl.,* **2011**, *50*(21), 4913-4917.
[http://dx.doi.org/10.1002/anie.201006870] [PMID: 21370365]

[60]   Shaikh, M.; Atyam, K.K.; Sahu, M.; Ranganath, K.V.S. Enhanced reactivity and selectivity of asymmetric *oxa*-Michael addition of 2′-hydroxychalcones in carbon confined spaces. *Chem. Commun. (Camb.),* **2017**, *53*(44), 6029-6032.
[http://dx.doi.org/10.1039/C7CC01096F] [PMID: 28524195]

[61]   Nonoyama, A.; Kumagi, N.; Shibasaki, M. Asymmetric flow catalysis: Mix-and- go solid phase Nd/Na catalyst for expedious enantioselective access to key intermediate of AZD7594. *Tetrahedron,* **2017**, *73*, 1517-1521.
[http://dx.doi.org/10.1016/j.tet.2017.01.066]

[62]   Ranganath, K.V.S.; Onitsuka, S.; Kiran Kumar, A.; Inanaga, J. Recent progress of N-heterocyclic carbenes in heterogeneous catalysis. *Catal. Sci. Technol.,* **2013**, *3*, 2161-2181.
[http://dx.doi.org/10.1039/c3cy00118k]

[63]   Kerr, M.S.; Read de Alaniz, J.; Rovis, T. An efficient synthesis of achiral and chiral 1,2,4-triazolium salts: bench stable precursors for N-heterocyclic carbenes. *J. Org. Chem.,* **2005**, *70*(14), 5725-5728.
[http://dx.doi.org/10.1021/jo050645n] [PMID: 15989360]

[64]   DiRocco, D.A.; Oberg, K.M.; Dalton, D.M.; Rovis, T. Catalytic asymmetric intermolecular stetter reaction of heterocyclic aldehydes with nitroalkenes: backbone fluorination improves selectivity. *J. Am. Chem. Soc.,* **2009**, *131*(31), 10872-10874.
[http://dx.doi.org/10.1021/ja904375q] [PMID: 19722669]

[65]   Ragno, D.; Carmine, G.D.; Brandolese, A.; Bortolini, O.; Giovannini, P.P.; Massi, A. Immobilization of privileged triazoliumcarbene catalyst for batch and flow stereo selective umpolung process. *ACS Catal.,* **2017**, *7*, 6365-6375.
[http://dx.doi.org/10.1021/acscatal.7b02164]

[66]   Oyamada, H.; Naito, T.; Kobayashi, S. Continuous flow hydrogenation using polysilane-supported palladium/alumina hybrid catalysts. *Beilstein J. Org. Chem.,* **2011**, *7*, 735-739.
[http://dx.doi.org/10.3762/bjoc.7.83] [PMID: 21804868]

[67]   Sheldon, R.A.; Arends, I.W.C.E.; Ten Brink, G.J.; Dijksman, A. Green, catalytic oxidations of alcohols. *Acc. Chem. Res.,* **2002**, *35*(9), 774-781.
[http://dx.doi.org/10.1021/ar010075n] [PMID: 12234207]

[68]   Shaikh, M.; Satanami, M.; Ranganath, K.V.S. Aerobic oxidation of benzoin using magnetite nanoparticles. *Catal. Commun.,* **2014**, *57*, 93-95.

[69]   Ballarin, B.; Barreca, D.; Boanini, E.; Cassani, M.C.; Dambruoso, P.; Massi, A.; Mignani, A.; Nanni, D.; Parise, C.; Zaghi, A. Supported gold nanoparticles for alcohols oxidation in continuous flow heterogeneous system. *ACS Sustain. Chem.& Eng.,* **2017**, *5*, 4746-4756.

[http://dx.doi.org/10.1021/acssuschemeng.7b00133]

[70] Porta, R.; Benaglia, M.; Chiroli, V.; Coccia, F.; Puglisi, A. Stereoselective Diels-Alder reactions promoted under continuous flow conditions by silica supported chiral organo catalysts. *Isr. J. Chem.,* **2014**, *54*, 381-394.
[http://dx.doi.org/10.1002/ijch.201300106]

[71] Burguete, M.I.; Cornejo, A.; Garcia-Verdugo, E.; Gil, M.J.; Luis, S.V.; Mayoral, J.A. M.-Merino, V.; Sokolava, M. Pybox monolithic mini flow reactors for continuous asymmetric cyclopropanation reaction under conventional and supercritical conditions. *J. Org. Chem.,* **2007**, *72*, 4344-4350.
[http://dx.doi.org/10.1021/jo070119r] [PMID: 17500566]

[72] Burguete, M.I.; Cornejo, A.; Garcia-Verdugo, E.; Gil, M.J.; Luis, S.V.; Mayoral, J.A. M.-Merino, V.; Sokolava, M. Bisoxazoline-functionalised enantioselective monolithic mini-flow-reactors: Development of efficient processes from batch to flow conditions. *Green Chem.,* **2007**, *9*, 1091-1096.
[http://dx.doi.org/10.1039/b704465h]

[73] Seayad, A.M.; Ramalingam, B.; Chai, C.L.L.; Li, C.; Garland, M.V.; Yoshinaga, K. Self-supported chiral titanium cluster (SCTC) as a robust catalyst for the asymmetric cyanation of imines under batch and continuous flow at room temperature. *Chemistry,* **2012**, *18*(18), 5693-5700.
[http://dx.doi.org/10.1002/chem.201200528] [PMID: 22438070]

[74] Alza, E.; Sayalero, S.; Cambeiro, X.C.; Martin, R.; Mirinda, P.O.; Pericàs, M.A. Catalytic batch and continuous flow production of highly enantioenriched cyclohexane derivatives with polymer-supported diarylprolinol silyl ethers. *Synlett,* **2011**, 464-468.

[75] Rasheed, M.; Elmore, S.C.; Wirth, T. *Asymmetric reactions in flow reactor: Catalytic Methods in Asymmetric Synthesis: Advanced Materials Techniques and Applications,* 1st ed.; John Wiley & Sons: New York, **2011**, pp. 345-371.

CHAPTER 4

Ball Milling: A Green Tool in Synthetic Organic Chemistry

Subhash Banerjee*, **Geetika Patel** and **Medha Kiran Patel**

Department of Chemistry, Guru Ghasidas Vishwavidyalaya (A Central University), Bilaspur 495009 (Chhattisgarh), India

Abstract: *Background:* Activation of covalent bonds for the initiation of chemical reactions can be achieved by all kinds of energy including light, thermal heating, microwave heating, electrical, sonochemical and mechanical energy. Among these, ball milling is an attractive alternative source of energy for the activation of bonds leading to chemical reactions due to its simplicity, ease of purification of products, mild reaction conditions and greenness of the process.

Methods: Mechano-chemical reaction is defined as "a chemical reaction that is induced by the direct absorption of mechanical energy." Simply, mechanical energy can be generated by grinding using a mortar and a pestle and the process of milling is carried out in ball mills. The process of milling is more reproducible due to the regulation of parameters like time and energy entry.

Results: The ball milling is mainly applicable in the industry for particle refinement processes, disagglomeration, the cracking of bacteria, *etc.* However, recently, ball milling has attracted considerable attention in organic synthesis due to its operational simplicity, economy, environment friendliness, and its potential to produce very good yields of products, and as a consequence, several research articles, review papers and book chapters have been published in recent time. The literature studies revealed that various carbon-carbon, carbon-heteroatom bond formation, condensation reactions, coupling reactions and oxidation-reduction reactions have been performed in a ball mill under mild and environmental-friendly reaction conditions.

Conclusion: The aim of this review is to highlight the recent breakthrough of ball milling in organic transformation leading to the synthesis of bioactive molecules in the context of *Green Chemistry.*

Keywords: Ball Milling, Clean Organic Synthesis, Green Synthetic Tools, Mechanochemical Reaction, Sustainable Alternative Energy Source.

---

* **Corresponding author Subhash Banerjee:** Department of Chemistry, Guru Ghasidas Vishwavidyalaya (A Central University), Bilaspur 495009 (Chhattisgarh), India; Tel: 07752-260203; E-mail: ocsb2006@gmail.com

Goutam Kumar Patra & Santosh Singh Thakur (Eds.)

INTRODUCTION OF ALL MILLING

General Introduction

The designing and development of a sustainable tool for the organic transformation can make an industrialized protocol economic, greener and more sustainable by reducing the energy, cost and radiation which are harmful to human health and the environment. Recently, several sustainable tools such as microwave irradiation, sonochemical, ball milling techniques have been developed for clean organic synthesis in the context of green chemistry. All these abovementioned techniques have their advantages and limitations. However, organic reactions using a mortar and pestle were well known since the early age of the progression of chemistry (Fig. **1a**). The process is known as a mechanochemical (grinding) reaction where energy is evolved during the grinding process. According to IUPAC, a mechanochemical reaction is referred to as "a chemical reaction that is induced by the direct absorption of mechanical energy" [1a]. On the other hand, grinding means subdividing the solids to finer products. However, the reactions initiated by so-called grinding process are very limited mainly due to inconsistent grinding speed and quite a low grinding strength. Over time with the development of technology, a mixer/shaker mill [1b] referred to as "milling" was introduced. The milling exerts higher energy and reliability than compared to grinding. Later on, "ball milling" has been developed when mechanical grinding has been added to those mills. Soon after development, the ball milling process has attracted tremendous attention to the community of material chemistry and synthetic organic chemistry [1c-h]. The ball milling has been established as one of the fastest-growing branches in the field of organic synthesis in the context of sustainable development due to the use of mechanochemical energy, solvent-free reaction conditions, *etc.* Moreover, the selectivity of the product formation could also be altered by using mechanochemistry as discussed by Bolm and Hernandez [1i]. Do *et al.* [1j] demonstrated significant achievements and opportunities created by mechanochemistry in the field of access to materials, molecular targets, and synthetic strategies which are unfeasible to access by conventional methods.

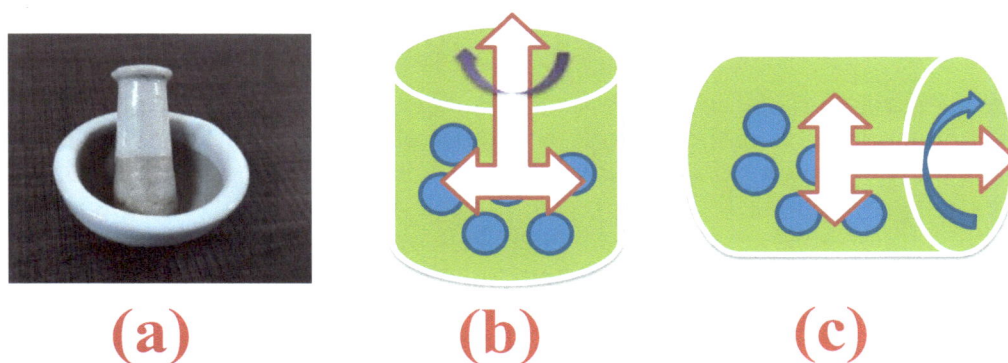

Fig. (1). (**a**) Represents mortar and pestle, (**b-c**) represent ball-milling.

Major Reaction Parameter in Ball Milling

There are various reaction parameters in the mechanochemical reactions using ball milling such as: (i) milling frequency and milling time; (ii) number and size of the ball used; and (iii) the material of milling balls and beakers. So far, mixer mill (MM) and planetary mill (PM) have been used frequently to perform organic transformations.

Principles of Ball Milling

The mechanism of reaction in ball mill is mainly based on the principle of crushing *i.e.* the reduction of the size of particles of the solid materials by collisions. A ball mill is made using a hollow cylindrical shell that can rotate horizontally or at a small angle to the horizontal about its axis. The horizontal shell is filled with grinding media, *i.e.* grinding ball of steel, stainless steel or rubber and the inner surface of the hollow cylindrical shell is usually lined with manganese steel or rubber-based scratch resistance material (Fig. **1b-c**). When cylindrical shell rotates, balls (occupying 30-50% of the volume of the ball mill) are lifted up and cascade down very frequently and by doing so size of the solid particles is reduced by the impact.

The process of reduction of size is very important and has wide applications. This process has several advantages such as (i) increase in the surface and consequently increase in the rate of physical and chemical processes, (ii) brings about excellent mixing of solids in solid-solid process, (iii) increases the solubility, (iv) increases dissolution rate, and (v) increases binding strength and dispersion properties of reactants.

ORGANIC REACTIONS UNDER BALL MILLING CONDITIONS

There are several unique and interesting advantages of ball milling as discussed in the introduction section and recently a number of chemical reactions including organic transformation have been investigated under ball milling conditions in the context of green chemistry. Very recently, the synthesis of organic compounds has been performed extensively under ball milling conditions due to its operational simplicity, economic, and environment-friendliness, and its potential to produce very good yields of products under mild reaction conditions. As a consequence, a number of research as letters, articles and reviews have been published in the literature [2]. In the next section, details of organic reactions under ball milling conditions have been presented.

Carbon-Carbon and Carbon-heteroatom Bond Formation *via* Michael Addition

Michael addition reaction is one of the best tools for the construction of carbon-carbon (C−C) and carbon-heteroatom (C−X; X is N, O, S, *etc.* heteroatoms) bonds. Michael addition is the addition of carbon nucleophiles (*e.g.,* active methylene compounds) to conjugated alkenes in the presence of a base or a Lewis acid. This reaction is referred to as classical Michael addition reaction. In the case of the nitrogen (*e.g.* amine), oxygen (*e.g.,* alcohol) and sulphur (*e.g.* thiol) nucleophiles, the reactions are referred to as aza-Michael, oxa-Michael and thia-Michael addition reactions, respectively. Most of the above-mentioned reactions are reported in solutions. However, over past decades, these reactions were performed under net conditions. The C−C and C−X bond formation reactions *via* Michael addition reactions under ball milling conditions have been discussed below.

Zhang and co-workers [3] have reported Michael addition of active methylene compounds to chalcone derivatives in the presence of K_2CO_3 as a mild base under ball milling conditions. They have observed that the Michael addition of ethyl acetoacetate and diethyl malonate to chalcone derivatives proceed smoothly with good yields of products (76-99%). The authors have reported that these reactions run very selectively in the presence of K_2CO_3 compared to other bases like $KF-Al_2O_3$ (Scheme **1**).

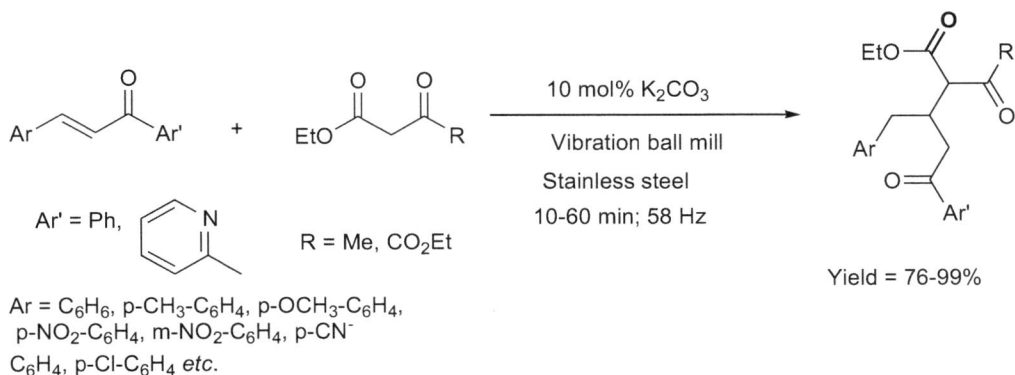

Scheme. (1).

The Zhang's ball milling method has several advantages over traditional methods such as solvent-free reaction conditions, shorter reaction time (10-60 min.), higher isolated yields (up to 99%) of the products, *etc.*

Kaupp *et al.* [4] introduced a catalyst-free methodology for the synthesis of oxygen- containing heterocycles *via* oxa-Michael addition followed by rearrangement and cyclization using a ball mill (Scheme **2**). In addition to the screening of chemical parameters, the authors have also investigated milling parameters such as ball diameter, milling time, number of balls, milling material, *etc.*

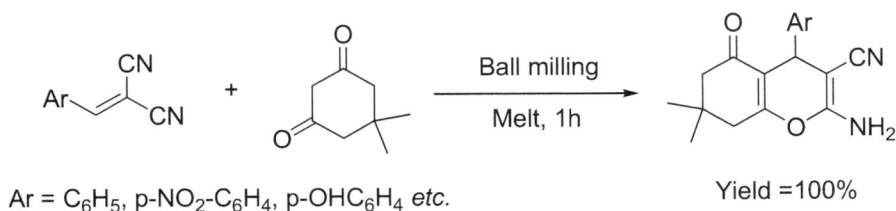

Scheme. (2).

Jin group's [5] showed that good yields of products were obtained when benzalacetophenone derivatives react with different amines without any catalyst under HSVM conditions at 30 Hz in 25 min. Significantly, reaction with (phenylmethyl)amine or hexahydropyridine gave good yields (up to 99%) without any side reaction in very shorter reaction time (Scheme **3**). They investigated various milling parameters such as milling time, frequency and milling materials, *etc.*

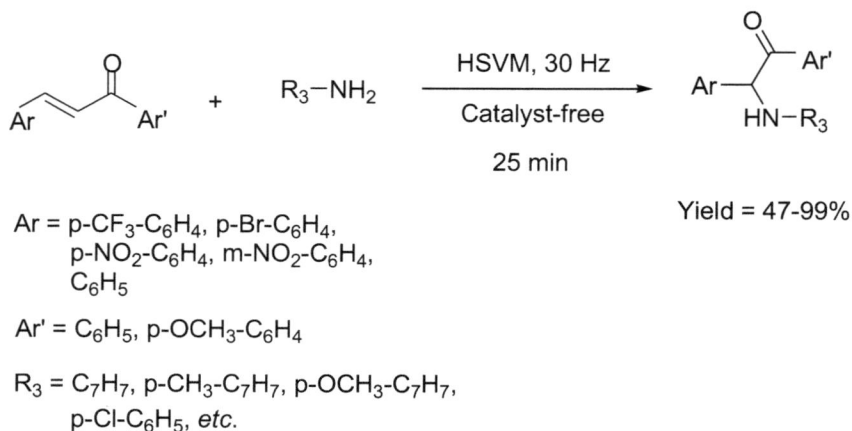

Ar = p-CF$_3$-C$_6$H$_4$, p-Br-C$_6$H$_4$,
      p-NO$_2$-C$_6$H$_4$, m-NO$_2$-C$_6$H$_4$,
      C$_6$H$_5$

Ar' = C$_6$H$_5$, p-OCH$_3$-C$_6$H$_4$

R$_3$ = C$_7$H$_7$, p-CH$_3$-C$_7$H$_7$, p-OCH$_3$-C$_7$H$_7$,
       p-Cl-C$_6$H$_5$, etc.

Yield = 47-99%

Scheme. (3).

Ge´rard *et al.* reported the synthesis of xanthone by the reaction of salicyladehyde and 2-cyclohexenone using a ball mill [6]. This is an example of oxa-Michael-aldol reaction (Scheme **4**).

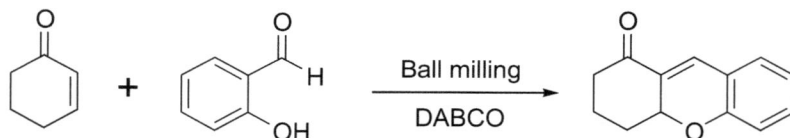

Scheme. (4).

Condensation Reactions

Elementary Condensation Reactions

Kaupp *et al.* [7] reported the synthesis of imines by the reactions of the equimolar amount of aldehydes and amines using a ball milling process under catalyst-free reaction conditions. The imines were obtained in 100% yields (Scheme **5**).

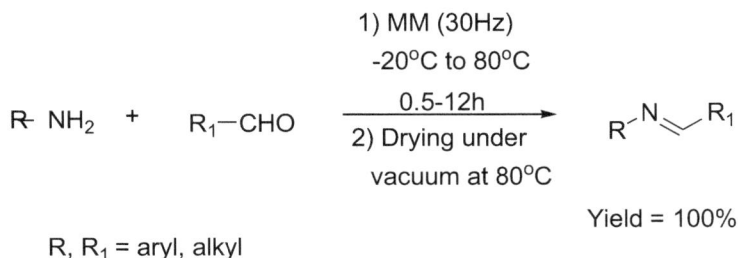

$$R-NH_2 \quad + \quad R_1-CHO \quad \xrightarrow[\substack{\text{2) Drying under} \\ \text{vacuum at 80°C}}]{\substack{\text{1) MM (30Hz)} \\ \text{-20°C to 80°C} \\ \text{0.5-12h}}} \quad R\text{-}N\text{=}R_1$$

Yield = 100%

R, R$_1$ = aryl, alkyl

Scheme. (5).

Naimi-Jamal and co-workers [8] investigated catalyst-free reactions of aldehydes or ketones with 2,4-dinitrophenyl hydrazine to produce 2,4-dinitrophenyl hydrazone derivatives by applying Kneading ball milling technique (Scheme **6**). The present method using a ball mill offers several advantages over traditional methods such as the use of solvent-free reaction conditions, shorter reaction time (10-20 min.), higher isolated yields (up to 100%), *etc.*

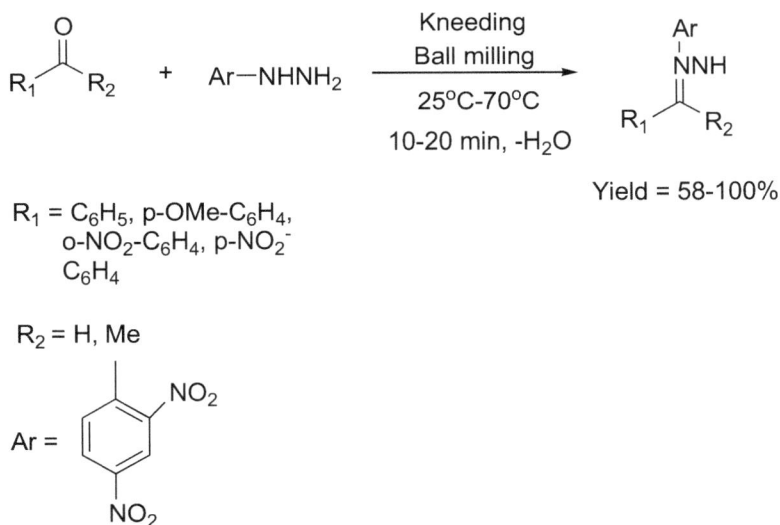

$$\underset{R_1 \quad R_2}{\overset{O}{\|}} \quad + \quad Ar-NHNH_2 \quad \xrightarrow[\substack{25°C\text{-}70°C \\ 10\text{-}20 \text{ min, } -H_2O}]{\substack{\text{Kneeding} \\ \text{Ball milling}}} \quad \underset{R_1 \quad R_2}{\overset{\substack{Ar \\ | \\ NNH}}{\|}}$$

Yield = 58-100%

R$_1$ = C$_6$H$_5$, p-OMe-C$_6$H$_4$,
    o-NO$_2$-C$_6$H$_4$, p-NO$_2^-$
    C$_6$H$_4$

R$_2$ = H, Me

Ar =

Scheme. (6).

Under solvent-free kneading ball milling, aldehydes or ketones react with hydroxylamine hydrochloride to produce oximes in excellent yields (75-100%) in the presence of aq. NaHCO$_3$ [8] (Scheme **7**). Shorter reaction time, excellent conversion and solvent-free reaction conditions are the major advantages of this method.

R$_1$ = C$_6$H$_5$, 4-OMe-C$_6$H$_4$, 2-NO$_2$-C$_6$H$_4$, 4-NO$_2$-C$_6$H$_4$

R$_2$ = H, Me

Scheme. (7).

Kaupp and co-workers [9] have reported a straight-forward and rapid method to access enamino ketones using a ball mill. The reactions between cyclic 1,3-dicarbonyl compounds and various anilines were conducted under catalyst-free conditions to produce excellent yields (99-100%) of products (Scheme **8**). The authors investigated the milling parameters such as ball diameter, milling time, number of balls, *etc.*

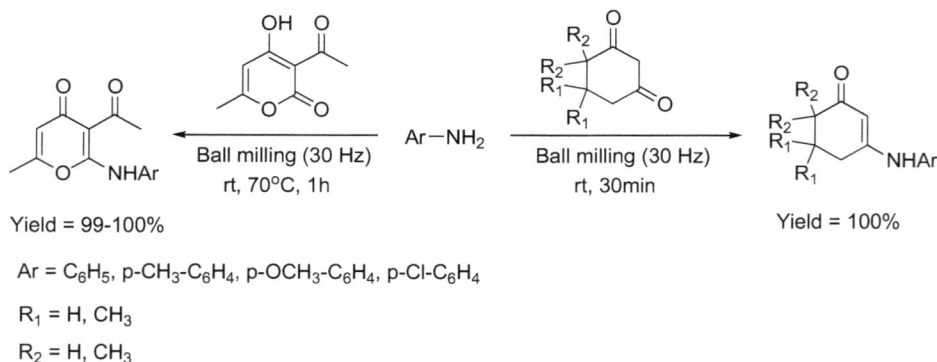

Yield = 99-100%

Ar = C$_6$H$_5$, p-CH$_3$-C$_6$H$_4$, p-OCH$_3$-C$_6$H$_4$, p-Cl-C$_6$H$_4$

R$_1$ = H, CH$_3$

R$_2$ = H, CH$_3$

Scheme. (8).

Lamaty *et al.* [10] reported the synthesis of nitrones by the condensation reactions of aldehydes with *N*-substituted hydroxylamines under ball milling conditions (Scheme **9**). They used four milling balls and silica as milling material.

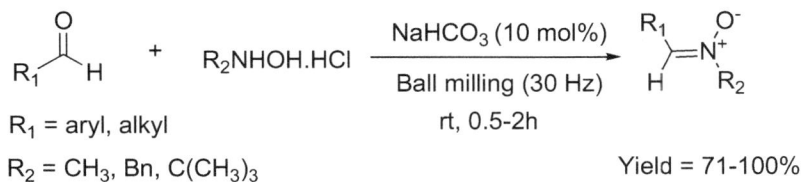

R$_1$ = aryl, alkyl

R$_2$ = CH$_3$, Bn, C(CH$_3$)$_3$

Yield = 71-100%

Scheme. (9).

Cyclocondensation-Heterocyclizataion Leading to Heterocyclic Compounds

Zhu and co-workers [11] reported the synthesis of flavone derivatives in good yields from 1-(2-hydroxyphenyl)-3-aryl-1,3-propanedione derivatives by using high-speed ball milling in the presence of KHSO$_4$ (Scheme **10**).

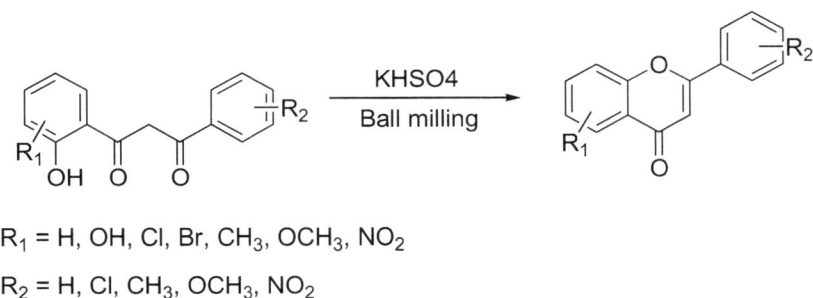

R$_1$ = H, OH, Cl, Br, CH$_3$, OCH$_3$, NO$_2$

R$_2$ = H, Cl, CH$_3$, OCH$_3$, NO$_2$

Scheme. (10).

Zhu group's [12] synthesized 1,3,5-triphenyl-4,5-dihydro-1H-pyrazoles by the reactions of chalcones and phenylhydrazine in the presence of NaHSO$_4$.H$_2$O in good yields (82-93%) under ball milling conditions at 21.5 Hz. The reactions were very fast (5-15 min) (Scheme **11**).

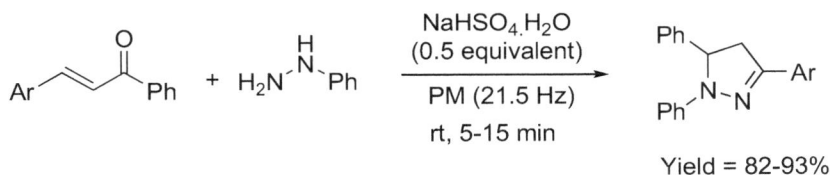

Yield = 82-93%

Scheme. (11).

Zhang and his group [13] demonstrated Diels-Alder cycloaddition reactions of ethenylbenzene and *N*-arylaldimines under ball milling conditions to produce good yields of *cis*-products (71-91%). The Diels-Alder cycloaddition reactions were carried out by using $FeCl_3$ in the absence of any organic solvent. The reactions were very fast and high yielding (Scheme **12**).

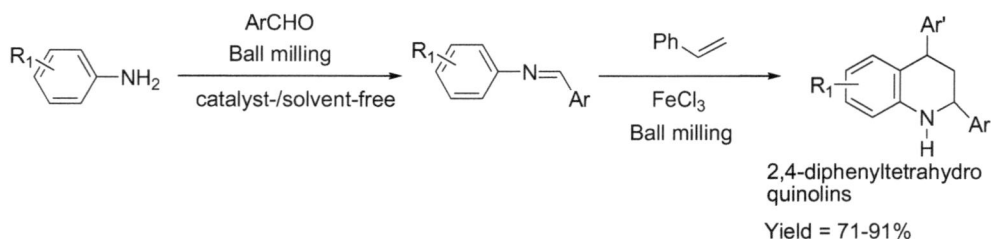

Scheme. (12).

Kaupp and co-workers [9] described how the mechanochemical reaction of stoichiometric (1:1 reactant ratio) amounts of thiourea derivatives and 2-bromoacetophenone afforded excellent yields (100%) of thiazole-hydrobromide derivatives without using any catalyst under ball milling conditions (Scheme **13**).

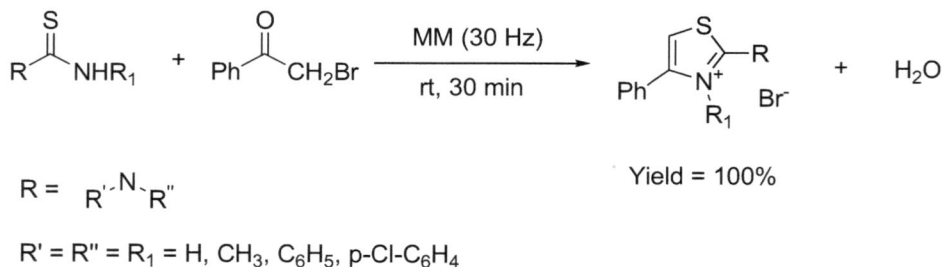

Scheme. (13).

Kaupp and co-workers [14a] reported a straight forward and facile route to access heterocycle products using a ball mill. The condensation reactions between 1,2-diaminobenzene and 1,2-dicarbonyl compounds took place smoothly to produce desired products in 100% yields without the production of byproducts (Scheme **14a**).

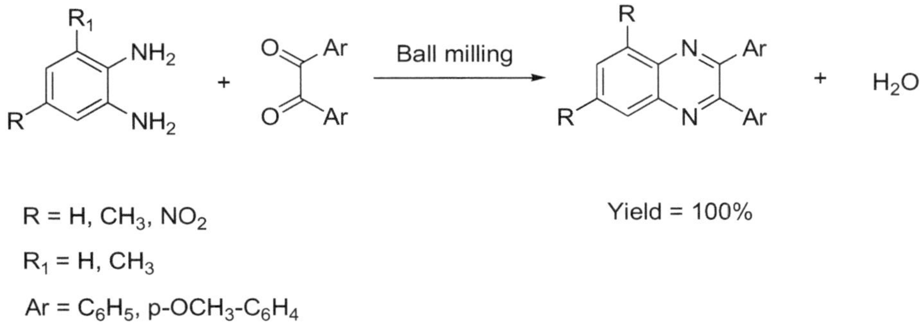

R = H, CH₃, NO₂

R₁ = H, CH₃

Ar = C₆H₅, p-OCH₃-C₆H₄

Yield = 100%

Scheme. (14a).

Jhang and his group [14b] reported a convenient solvent-free method for the synthesis of benzothiazole, benzimidazole and benzoxazole derivatives using ZnO NPs as a recyclable catalyst under ball-milling conditions. This method using a ball mill afforded excellent yields under environmentally-friendly conditions. This process is also equally applicable to multi-gram scale synthesis (Scheme **14b**).

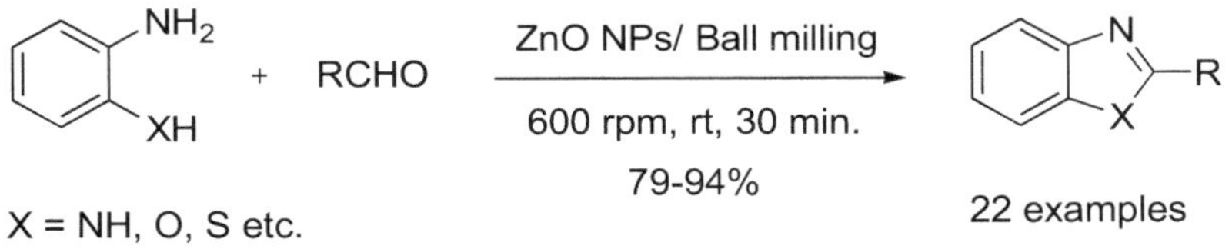

X = NH, O, S etc.

Scheme. (14b).

A series of quinoxalines were synthesized by Oliveira and co-workers [15] in good yields by condensation reactions of 1,2-diaminoarenes with 1,2-dicarbonyl compounds under ball mill conditions (Scheme **15**).

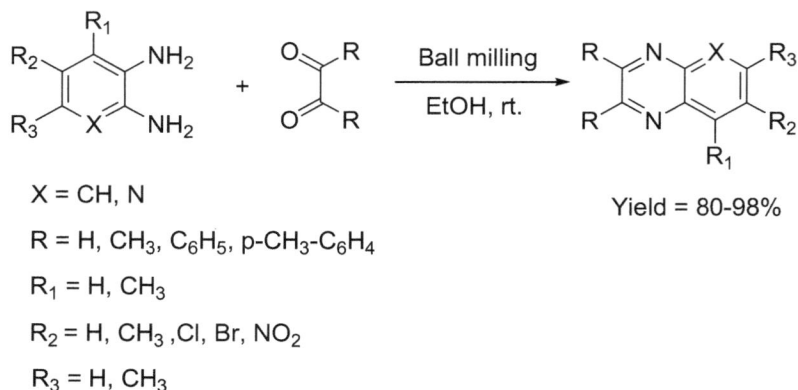

X = CH, N

R = H, CH$_3$, C$_6$H$_5$, p-CH$_3$-C$_6$H$_4$

R$_1$ = H, CH$_3$

R$_2$ = H, CH$_3$,Cl, Br, NO$_2$

R$_3$ = H, CH$_3$

Yield = 80-98%

Scheme. (15).

Carlier and co-workers [16] performed the reactions between o-diaminoarenes and 1,2- or 1,3-dicarbonyls to synthesize dibenzophenazines and dibenzopyridoquinoxaline derivatives, respectively. These products were obtained in good yields under solvent and catalyst-free reaction conditions (Scheme **16**).

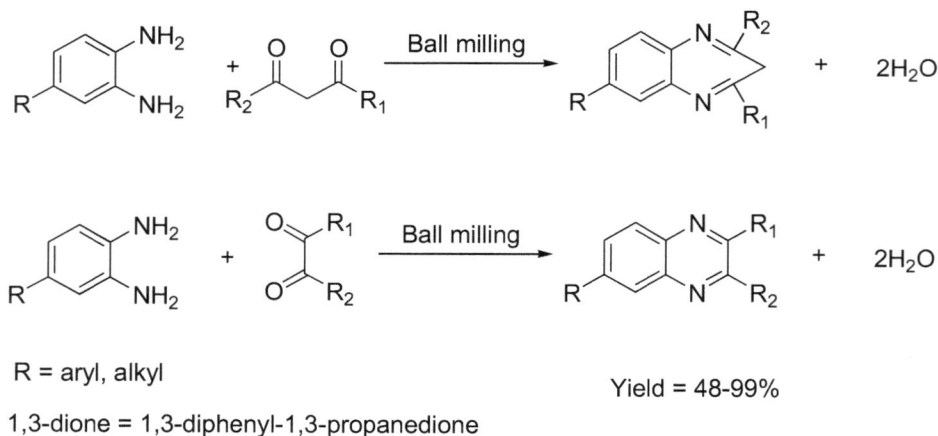

R = aryl, alkyl

1,3-dione = 1,3-diphenyl-1,3-propanedione

Yield = 48-99%

Scheme. (16).

Zeng and co-workers [17] reported the synthesis of pyrrole derivatives from amines and ketone or aldehydes derivatives in good yields without using any solvent by using high- speed ball milling. The authors used Mn(OAc)$_3$ as a mediator (Scheme **17**).

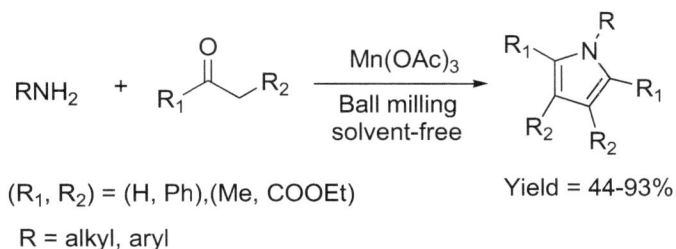

(R_1, R_2) = (H, Ph),(Me, COOEt)   Yield = 44-93%

R = alkyl, aryl

Scheme. (17).

Kaupp *et al.* [18] demonstrated solid-solid reactions of aniline derivatives with *O*-phenylenediisothiocyanate in the absence of a catalyst under mechanochemical conditions to produce substituted (anilino-thiocarbonyl)-benzimidazolidin--2-thiones in 100% yields (Scheme **18**). The authors have carried out the reactions at 50 Hz milling frequency.

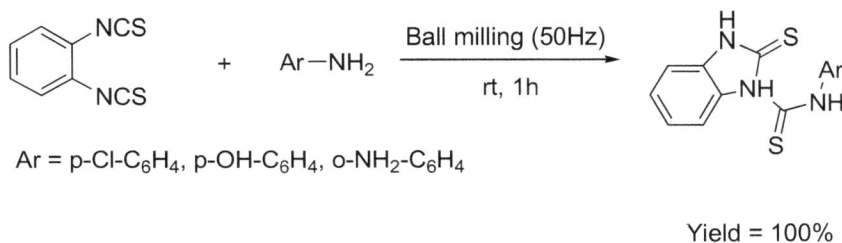

Ar = p-Cl-C_6H_4, p-OH-C_6H_4, o-NH_2-C_6H_4

Yield = 100%

Scheme. (18).

Cyclocondensation Reactions

Brase and co-workers [6] introduced chemoselective base-catalyzed condensation of salicaldehyde with 1,2−unsaturated aldehydes in a ball mill. The reactions, in the presence of base DABCO, gave chromene derivative as a major product but in the presence of Et_3N produced dihydrobenzene pyran as the major product. To obtain the best conditions for these reactions, the author studied the relationships among rotational frequency, ball weight, and time of reactivity during milling (Scheme **19**).

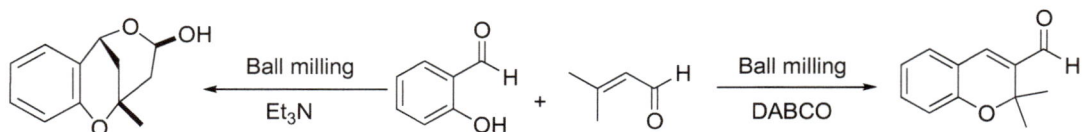

Scheme. (19).

Kaupp *et al.* [19] developed a solvent-free methodology for the synthesis of (R)-thiazole quantitative yield involving reaction of L-cysteine and formaldehyde in a vacuum under ball milling condition at 80 °C. The authors have used four balls (Scheme **20**).

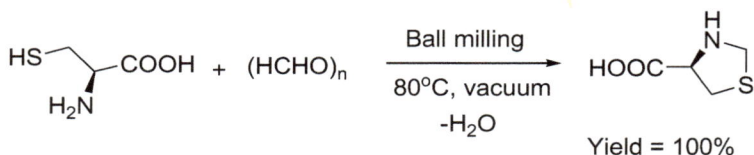

Scheme. (20).

Kaupp *et al.* [9] again demonstrated solid-state mechanochemical cascade condensation reactions. The 1,2-diaminobenzene reacts with ketoglutaric acid/or ethanedioic acid/or 1,4-dihydroquinoxaline-2,3-dione to give different products as shown in Scheme (**21**) in good to excellent yields (90-100%).

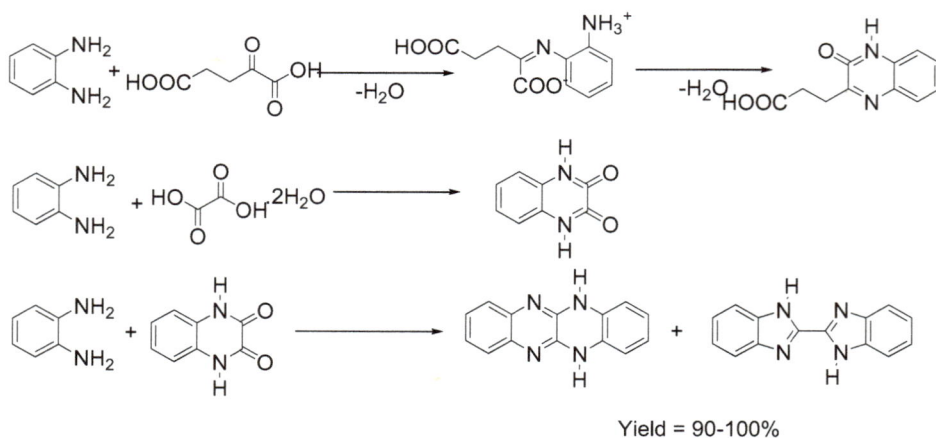

Scheme. (21).

Condensation Cycloaddition Reactions

Paveglio and co-workers [20] investigated the synthesis of 1-*H*-pyrazole derivatives in a ball mill using *p*-toluenesulfonic acid (10 mol%) as a catalyst (Scheme **22**). They found the optimum yield of the products at a frequency of 450 rpm. The reactions were very fast (3 min.) and produced high yields.

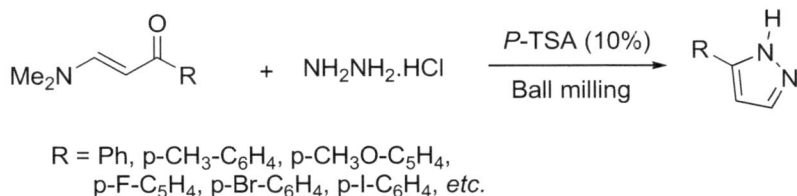

R = Ph, p-CH$_3$-C$_6$H$_4$, p-CH$_3$O-C$_5$H$_4$,
p-F-C$_5$H$_4$, p-Br-C$_6$H$_4$, p-I-C$_6$H$_4$, *etc.*

Scheme. (22).

Zhang and co-workers [21] reported one-pot method for the synthesis 3,5-diphenyl-1*H*-pyrazoles using sodium persulfate as an oxidant under solvent-free ball milling conditions. The use of mechanochemical energy, higher yields (80-93%), and solvent-free protocol are the major advantages of the present protocol (Scheme **23**).

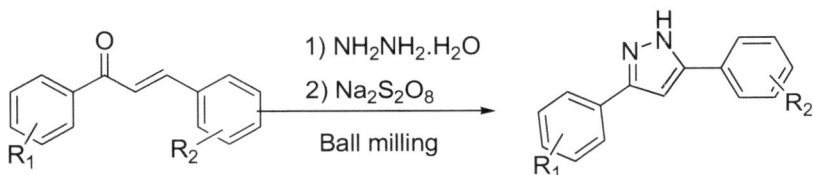

Scheme. (23).

A green synthetic method for the preparation of 2,3,6,7-tetrahydro-4(5*H*)benzofuranone derivatives by using Mn(OAc)$_3$.2H$_2$O has been reported by Wang *et al.* [22] Here, Mn(OAc)$_3$.2H$_2$O acts as an oxidant as well as Lewis acid. The reactions were performed in the absence of solvent under ball milling conditions at 30 Hz frequency for 1h (Scheme **24**).

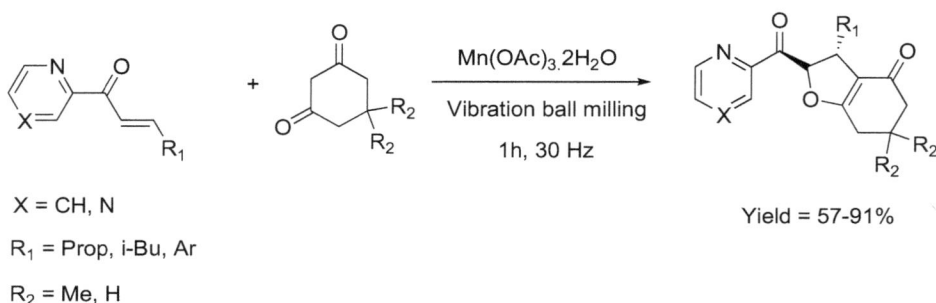

X = CH, N

R$_1$ = Prop, i-Bu, Ar

R$_2$ = Me, H

Yield = 57-91%

Scheme. (24).

A simple and efficient protocol for the synthesis of arylbenzodioxyxanthenedione scaffolds has been developed by Banerjee and his group [23a] *via* one-pot multi-component reactions of aromatic aldehydes, 2-hydroxy-1,4-naphthoquinone and 3,4-methylenedioxy phenol using mesoporous PbO nanoparticles as a catalyst under ball milling conditions. The reported method offered several advantages like shorter reaction time (60 min.), excellent yields of products (92-97%), solvent-free conditions, use of mild and reusable PbO NPs as a catalyst, simple purification of products and finally use of green ball milling technique. They have observed that at 600 rpm using six 6 balls of 10 mm diameter produced maximum yields of the products at room temperature (Scheme **25a**).

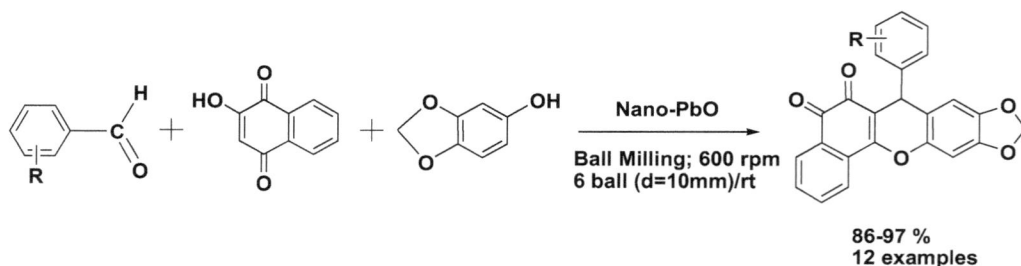

86-97 %
12 examples

Scheme. (25a).

Banerjee *et al.* [23b] again reported the synthesis of 4-oxo-tetrahydroindoles using sulfamic acid under ball milling conditions. The reaction was performed under solvent-free conditions. The milling was performed at 600 rpm using six balls of 10 mm diameter (Scheme **25b**).

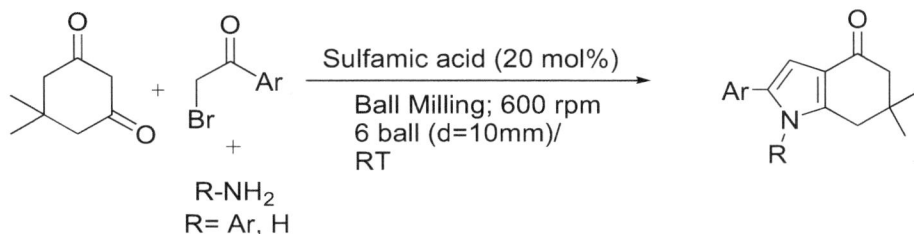

Scheme. (25b).

Cyclization Addition *via* Click Reactionss

Stolle *et al.* [24] developed a green method for the synthesis of triazole derivatives in quantitative yields (84%-91%) by reacting equimolar amounts of azides and alkynes in the presence of $Cu(OAc)_2$ (5 mol%). They have used SiO_2 (5g) as milling material, and 13.3 Hz milling frequency. All the reactions were completed within 10 min. The solvent-free reaction condition is another major advantage of the present ball milling process (Scheme **26**).

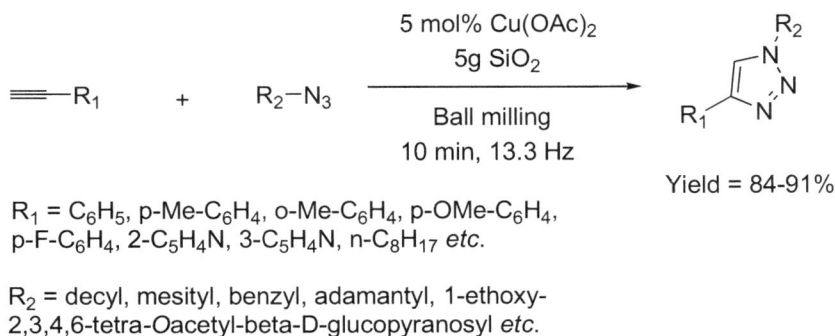

$R_1 = C_6H_5$, p-Me-C_6H_4, o-Me-C_6H_4, p-OMe-C_6H_4, p-F-C_6H_4, 2-C_5H_4N, 3-C_5H_4N, n-C_8H_{17} *etc.*

R_2 = decyl, mesityl, benzyl, adamantyl, 1-ethoxy-2,3,4,6-tetra-Oacetyl-beta-D-glucopyranosyl *etc.*

Scheme. (26).

Under mechanical milling conditions, polymerization of 1,12-diazidododecane and bis-ethynyl compounds was performed in the presence of a catalytic amount of $Cu(OAc)_2$ (Scheme **27**) [24].

R = n-C$_4$H$_8$, 1,4-C$_6$H$_2$-2,5-OC$_8$H$_{17}$

Scheme. (27).

Ranu and co-workers [25] reported a solvent-free one-pot three-component reaction of alkyl halide/aryl boronic acid, sodium azide and terminal alkynes under ball milling conditions on the surface of Cu/Al$_2$O$_3$ produced 1,4-disubstituted-1,2,3-triazoles (Scheme **28**). The authors reported that the reaction neither requires any toxic organic solvent nor any separation/purification and simply gives aryl-alkyl and aryl-aryl substituted 1,2,3-triazoles. The optimum frequency required for this reaction is reported to be 600 rpm.

R$_1$ = alkyl; R$_2$ = alkyl, aryl; R$_3$ = aryl, alkenyl, heteroaryl etc

Scheme. (28).

Strukil and co-workers [26] reported the synthesis of symmetrical and non-symmetrical phenylenediamine(thio)urea derivatives by using a ball mill in excellent yields under organic solvent-free conditions (Scheme **29**).

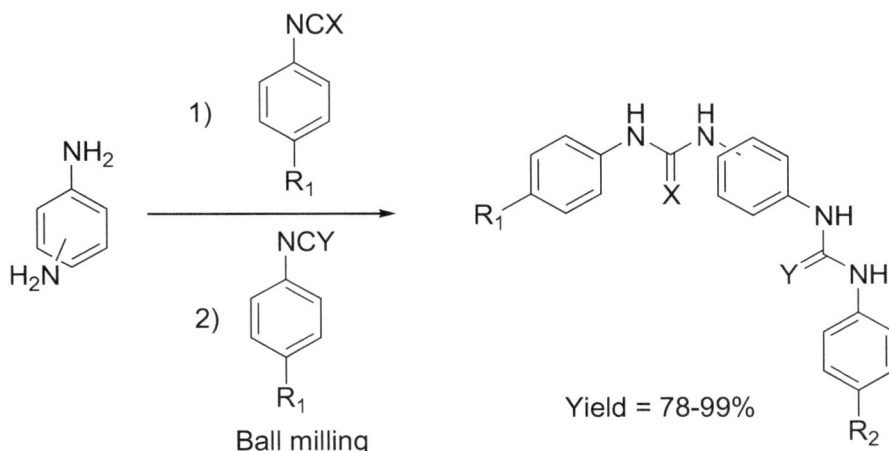

Scheme. (29).

Miscellaneous Reactions

Carbon- Heteroatom Bond Formation via Substitution Reactions

Mack and co-workers [27] reported different reactions of 4-bromobenzyl bromide with metal halides in the presence or absence of 18-crown-6-ether. The author reported that in the absence of 18-crown-6-ether by using ball milling (frequency = 50 Hz) reactions between 4-bromobenzyl bromide and MX (M = Na, K, Cs; X=F, Cl, I, SCN, N$_3$etc.) give low to good yields (14-97%) of products. The author also reported that the reaction between 4-bromobenzyl bromide and KX (X= -F, -Cl, -I, -CN, -OAc) in the presence of 18-crown-6-ether by using ball milling (50 Hz) gave 70-90% yields. Excellent yields of products were obtained in the presence of 18-crown-6-ether due to the formation of a complex which increases the basicity and nucleophilicity of nucleophiles in nucleophilic addition reactions (Scheme **30**).

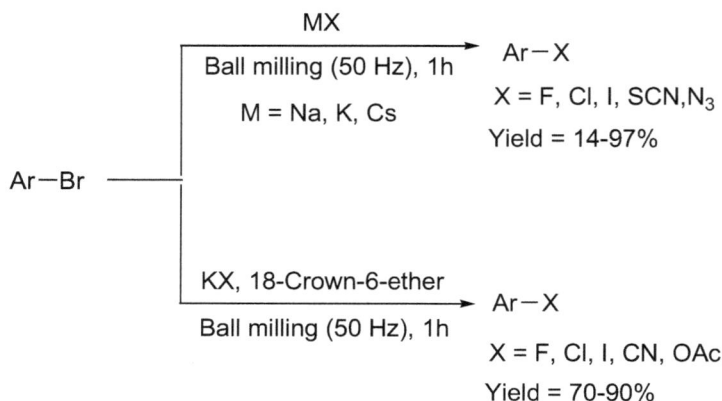

Scheme. (30).

Carbon-Heteroatom Bond Formation via Addition Followed by Cycloaddition Reactions

Phung *et al.* [28] investigated catalyst–and solvent-free reactions of 2-alkyl or 2-aryl aziridine with CO_2 by applying ball milling conditions at 18 Hz to produce differently substituted oxazolidinones (Scheme **31**). However, longer reaction time (17 h) is one of the major limitations of these methods.

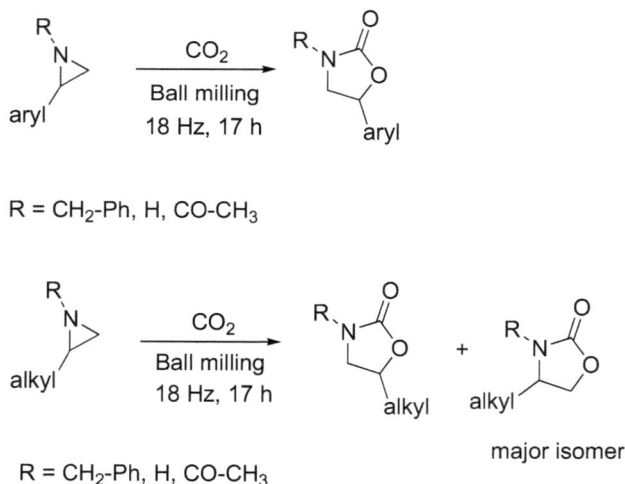

R = CH_2-Ph, H, CO-CH_3

R = CH_2-Ph, H, CO-CH_3

Scheme. (31).

Stolle and co-workers [29] reported the mechanochemical synthesis of enamine in good yields (81-94%) by reacting amines derivatives with dialkyl butynedioate without using any solvent. The reactions produced both *E* and *Z* isomers. The reactions were carried out on the surface of SiO_2 which plays the role of a grinding auxiliary (Scheme **32**).

R$_1$ = CO$_2$Me, CO$_2$Et

R$_2$ = H, CO$_2$Me, CO$_2$Et

R$_3$ = R$_4$ = H, Ph, p-CH$_3$-C$_6$H$_5$, ^{i}Pr

Scheme. (32).

Coupling Reactions

Zille and co-workers [30] reported the synthesis of indole derivatives in quantitative yields under solvent mechanochemical conditions. The reactions were performed by reacting terminal alkynes and 2-iodoaniline (Sonogashira reaction) catalyzed by $ZnBr_2$ using high-speed ball milling (Scheme **33**).

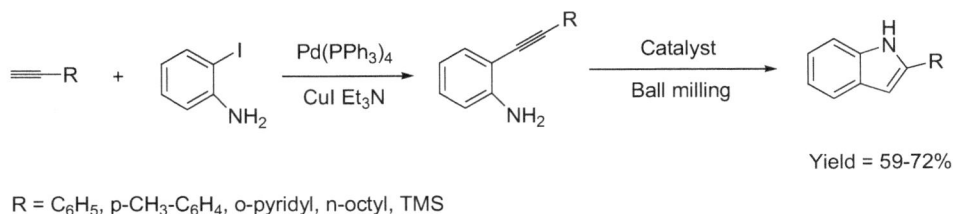

R = C$_6$H$_5$, p-CH$_3$-C$_6$H$_4$, o-pyridyl, n-octyl, TMS

Scheme. (33).

Naphthopyran derivatives were synthesized in good yields (76-97%) by cycloaddition reactions of ethynylcarbinol derivatives with 2-naphthol a catalytic amount of $InCl_3.4H_2O$ using ball mill conditions by Dong and his group [31]. Here, $InCl_3.4H_2O$ promoted the reaction by acting as Lewis acid. The reactions were performed in 30 Hz frequency for 1 h (Scheme **34**).

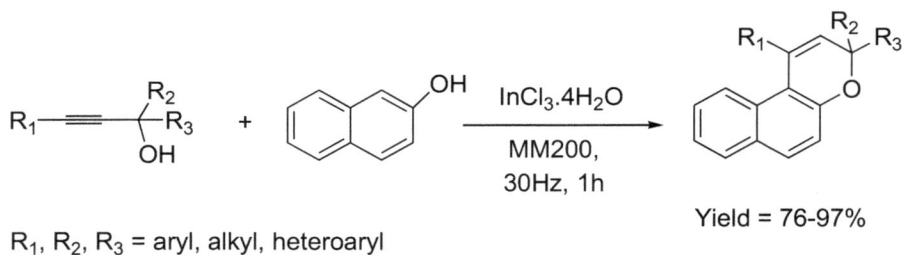

R$_1$, R$_2$, R$_3$ = aryl, alkyl, heteroaryl

Scheme. (34).

Carbon-Carbon and Carbon-heteroatom Bond Formation via Alkylation Reactions

Kaupp *et al.* [9] performed a solid-solid reaction of bromoacetophenone with benzimidazolethiol in the absence of catalyst, under mechanochemical conditions at a frequency of 30 Hz to produce thiouronium salt in quantitative yield (100%) (Scheme **35**).

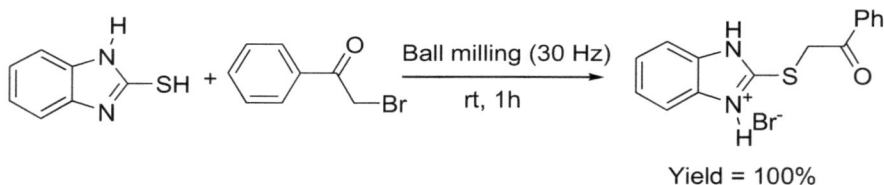

Yield = 100%

Scheme. (35).

Mack *et al.* [32] introduced ball milling techniques for the alkylation of alkali salts with p-bromobenzyl bromides. The reactions were performed in the presence of 18-crown-6, by applying milling frequency of 17.7 Hz and produced good yields of the products (Scheme **36**).

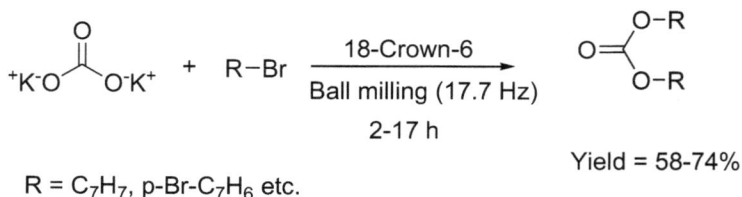

R = C$_7$H$_7$, p-Br-C$_7$H$_6$ etc.

Scheme. (36).

E. Abdel-Latif *et al.* [33] developed a simple and efficient protocol for the synthesis of pyrazolyl ether derivatives by using a ball mill. They performed reactions at 20-25 Hz milling frequency at room temperature and pyrazolyl ether derivatives were obtained in 100% yields after drying at 80 °C in a vacuum (Scheme **37**).

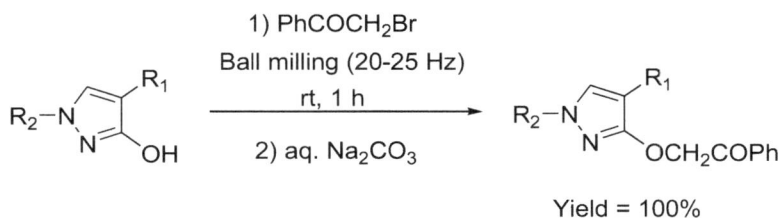

Yield = 100%

Scheme. (37).

Carbon-Nitrogen Bond Formation via Carbon-Hydrogen Bond Activation

Wang and Gao [34] have investigated the reactions between aldehydes and anilines using under mechanochemical conditions under solvent-free conditions. They have performed the reactions at milling frequency of 30 Hz using oxone as an oxidant, $MgSO_4$ dehydrating agent and (Scheme **38**).

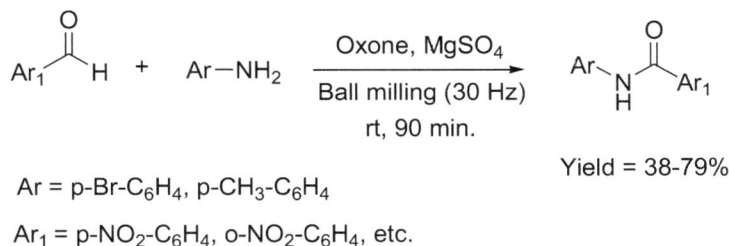

Yield = 38-79%

Ar = p-Br-C_6H_4, p-CH$_3$-C_6H_4

Ar$_1$ = p-NO$_2$-C_6H_4, o-NO$_2$-C_6H_4, etc.

Scheme. (38).

Transesterification Reactions under Ball Milling Conditions

Ranu and co-workers [35] have reported trans-esterification reactions under solvent-free conditions using basic Al_2O_3 (3.5 g) as milling material at milling frequency of 600 rpm. The advantage of this method is that, a variety of synthesized products were obtained with various esters and alcohols under mild reaction conditions in the absence of protic acid (Scheme **39**).

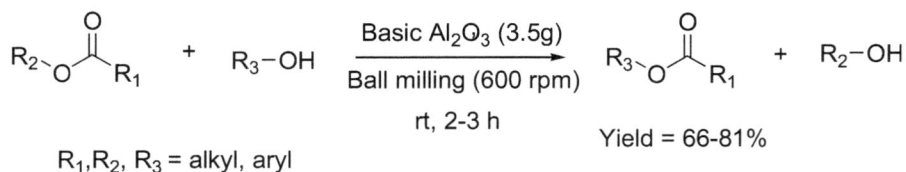

R_1, R_2, R_3 = alkyl, aryl

Scheme. (39).

Oxidation and Reduction Reactions

Thorwirth *et al.* [36] have reported a solvent-free method for the chemoselective oxidation of anilines to the corresponding azo or azoxy homocoupling products in a planetary ball mill using $KMnO_4$ as oxidant (Scheme **40**). The authors have screened different milling parameters to optimize reaction conditions. They have investigated the effect of different milling auxiliaries like silica gel, alumina, montmorillonite, etc and alumina (activity 90) was proven to be the best choice of auxiliary. In addition, the reaction using six balls of 15 mm diameter at 800 rpm for ~10 min was established as the best choice for getting optimum yields.

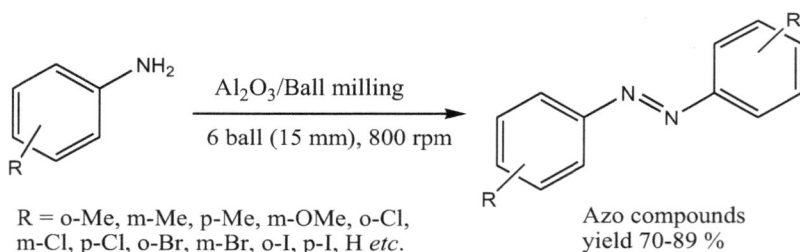

R = o-Me, m-Me, p-Me, m-OMe, o-Cl,   Azo compounds
m-Cl, p-Cl, o-Br, m-Br, o-I, p-I, H *etc.*   yield 70-89 %

Scheme. (40).

Szuppa *et al.* [37] have demonstrated oxidation of γ-terpenesto *p*-cymene under planetary ball milling conditions using $KMnO_4$ as oxidant (Scheme **41**). They have observed that alumina is the best as a milling material.

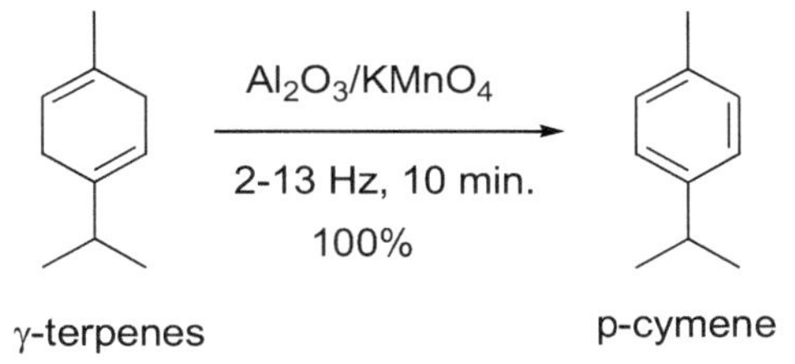

γ-terpenes   p-cymene

Scheme. (41).

Szuppaand co-workers [38] have reported oxidation of β-pinene to nopinone using alumina- $KMnO_4$ as oxidant under ball milling conditions (Scheme **42**).

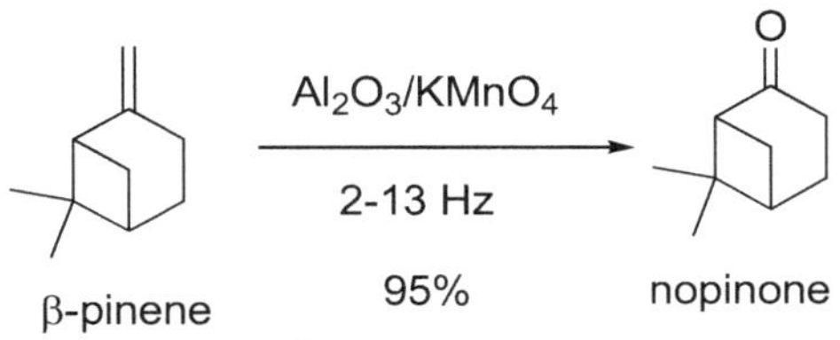

β-pinene   95%   nopinone
Scheme 42

Scheme. (42).

Mack *et al.* [39] have reported solid-state reduction of aromatic carbonyl compounds to the corresponding benzylic alcohols (32-85%) by using ball milling conditions for 1-17 h (Scheme **43**). They have used Al_2O_3 as milling material and found that 17 Hz is the optimum frequency required for this conversion.

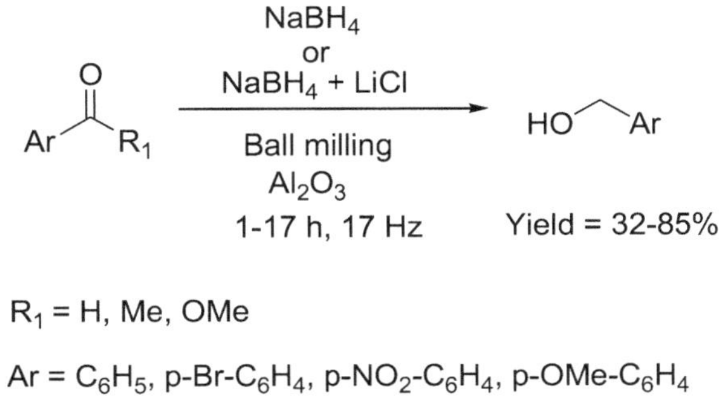

R_1 = H, Me, OMe

Ar = C_6H_5, p-Br-C_6H_4, p-NO_2-C_6H_4, p-OMe-C_6H_4

Scheme. (43).

CONCLUSION

In this chapter, we have discussed the detail applications of ball milling in the carbon-carbon and carbon-heteroatom bond formation *via* Michael addition, alkylation and cycloaddition, and condensation reaction leading to the synthesis of heterocyclic molecules of biological importance. All the organic transformations presented in this chapter have been performed under ball milling conditions and have several advantages. Most of the advantages fullfil many aspects of green chemistry, like the use of mechanochemical energy as the source of energy, solvent-free and mild reaction conditions, and good to excellent yields of products obtained with high purity even without column chromatography. Moreover, ball milling has proved to be a greener alternative to other energy sources like conventional heating, microwave irradiation, sonication, *etc.* and thus will find many more applications in organic synthesis in the near future.

CONSENT FOR PUBLICATION

Not applicable.

CONFLICT OF INTEREST

The author declares that there is no conflict of interest in this chapter.

ACKNOWLEDGEMENTS

Declared none.

REFERENCES

[1]   a) McNaught, A.D.; Wilkinson, A. *Compendium of Chemical Terminology; IUPAC Recommendations*; IUPAC: Zürich, Switzerland, **1979**. b) Ranu, B.C.; Chatterjee, T.; Mukherjee, N. Carbon–Heteroatom Bond Forming Reactions and Heterocycle Synthesis under Ball Milling. In: *Ball Milling Towards Green Synthesis: Applications, Projects*; Challenges, **2014**; pp. 1-33.
[http://dx.doi.org/10.1039/9781782621980-00001] c) Soori, F.; Nezamzadeh-Ejhieh, A. Synergistic effects of copper oxide-zeolite nanoparticles composite on photocatalytic degradation of 2,6-dimethylphenol aqueous solution. *J. Mol. Liq.,* **2018**, *255*, 250-256.
[http://dx.doi.org/10.1016/j.molliq.2018.01.169] d) Nezamzadeh-Ejhieh, A.; Shirzadi, A. Enhancement of the photocatalytic activity of Ferrous Oxide by doping onto the nano-clinoptilolite particles towards photodegradation of tetracycline. *Chemosphere,* **2014**, *107*, 136-144.
[http://dx.doi.org/10.1016/j.chemosphere.2014.02.015] [PMID: 24875881] e) Nezamzadeh-Ejhieh, A.; Moeinirad, S. Heterogeneous photocatalyticdegradation of furfural using NiS-clinoptilolite zeolite. *Desalination,* **2011**, *273*, 248-257.
[http://dx.doi.org/10.1016/j.desal.2010.12.031] f) Nezamzadeh-Ejhieh, A.; Khodabakhshi-Chermahini, F. Incorporated ZnO onto nanoclinoptilolite particles as the active centers in the photodegradation of phenylhydrazine. *J. Ind. Eng. Chem.,* **2014**, *20*, 695-704.
[http://dx.doi.org/10.1016/j.jiec.2013.05.035] g) Nezamzadeh-Ejhieh, A.; Shahanshahi, M. Modification of clinoptilolitenano-particles with hexadecylpyridynium bromide surfactant as an active component of Cr(VI) selective electrode. *J. Ind. Eng. Chem.,* **2013**, *19*, 2026-2033.
[http://dx.doi.org/10.1016/j.jiec.2013.03.018] h) Hasheminejad, M.; Nezamzadeh-Ejhieh, A. A novel

citrate selective electrode based on surfactant modified nano-clinoptilolite. *Food Chem.,* **2015**, *172*, 794-801.
[http://dx.doi.org/10.1016/j.foodchem.2014.09.057] [PMID: 25442622] i) Hernández, J.G.; Bolm, C. Altering Product Selectivity by Mechanochemistry. *J. Org. Chem.,* **2017**, *82*(8), 4007-4019.
[http://dx.doi.org/10.1021/acs.joc.6b02887] [PMID: 28080050] j) Do, J-L.; Friščić, T. Mechanochemistry: A Force of Synthesis. *ACS Cent. Sci.,* **2017**, *3*(1), 13-19.
[http://dx.doi.org/10.1021/acscentsci.6b00277] [PMID: 28149948]

[2]   a) OuldM'hamed, M. Ball milling for heterocyclic compounds synthesis in green chemistry. *Synth. Commun.,* **2015**, *45*, 2511-2528.
[http://dx.doi.org/10.1080/00397911.2015.1058396] b) Stolle, A.; Szuppa, T.; Leonhardt, S.E.; Ondruschka, B. Ball milling in organic synthesis: solutions and challenges. *Chem. Soc. Rev.,* **2011**, *40*(5), 2317-2329.
[http://dx.doi.org/10.1039/c0cs00195c] [PMID: 21387034] c) Wang, G-W. Mechanochemical organic synthesis. *Chem. Soc. Rev.,* **2013**, *42*(18), 7668-7700.
[http://dx.doi.org/10.1039/c3cs35526h] [PMID: 23660585] Margetic, D.; Štrukil, V. *Mechanochemical Organic Synthesis*; Elsevier New York: NY, USA, **2016**. e) Friščić, T. Supramolecular concepts and new techniques in mechanochemistry: cocrystals, cages, rotaxanes, open metal-organic frameworks. *Chem. Soc. Rev.,* **2012**, *41*(9), 3493-3510.
[http://dx.doi.org/10.1039/c2cs15332g] [PMID: 22371100] f) El-Sayed, T.H.; Aboelnaga, A.; El-Atawy, M.A.; Hagar, M. Ball Milling Promoted *N*-Heterocycles Synthesis. *Molecules,* **2018**, *23*(6), 1348.
[http://dx.doi.org/10.3390/molecules23061348] [PMID: 29867039]

[3]   a) Zhang, Z.; Dong, Y-W.; Wang, G-W.; Komatsu, K. Highly efficient mechanochemical reactions of 1,3-Dicarbonyl compounds with chalcones and azachalconesCatalyzed by potassium carbonate. *Synlett,* **2004**, 61-64.b) Zhang, Z.; Dong, Y-W.; Wang, G-W.; Komatsu, K. Mechanochemical Michael Reactions of Chalcones and Azachalcones with Ethyl Acetoacetate Catalyzed by K_2CO_3 under Solvent-Free Conditions. *Chem. Lett.,* **2004**, *33*, 168-169.
[http://dx.doi.org/10.1246/cl.2004.168]

[4]   Kaupp, G.; Naimi-Jamal, M.R.; Schmeyers, J. Solvent-free knoevenagel condensations and michael additions in the solid state and in the melt with quantitative yield. *Tetrahedron,* **2003**, *59*(21), 3753-3760.
[http://dx.doi.org/10.1016/S0040-4020(03)00554-4]

[5]   Jin, L.Y.; Wen, C.Y.; Shuang, X.F.; Ming, F.W.; Bin, Y.W.; Hong, J.J.; Rong, G.J. Solvent and Catalyst Free Azo-Michael Addition under High-Speed Vibration Milling. *Sci. China Chem.,* **2012**, *55*, 1252-1256.
[http://dx.doi.org/10.1007/s11426-012-4605-y]

[6]   Ge'rard, E.M.C.; Sahin, H.; Encinas, A.; Bra¨se, S. Systematic Study of a Solvent-Free Mechanochemically Induced Domino Oxa-Michael-Aldol Reaction in a Ball Mill. *Synlett,* **2008**, 2702-2704.

[7]   a) Schmeyers, J.; Toda, F.; Boy, J. andKaupp, G. Quantitative solid–solid synthesis of azomethines. *J. Chem. Soc., Perkin Trans.,* **1998**, *2*, 989-994.
[http://dx.doi.org/10.1039/a704633b] b) Kaupp, G.; Schmeyers, J.; Boy, J. Waste-free solid-state syntheses with quantitative yield. *Chemosphere,* **2001**, *43*(1), 55-61.
[http://dx.doi.org/10.1016/S0045-6535(00)00324-6] [PMID: 11233826]

[8]   Mokhtari, J.; Naimi-Jamal, M.R.; Hamzeali, H.; Dekamin, M.G.; Kaupp, G. Kneading ball-milling and stoichiometric melts for the quantitative derivatization of carbonyl compounds with gas-solid recovery. *ChemSusChem,* **2009**, *2*(3), 248-254.
[http://dx.doi.org/10.1002/cssc.200800258] [PMID: 19266517]

[9]   Kaupp, G.; Schmeyers, J.; Boy, J. Iminium Salts in Solid-State Synthesis Giving 100% Yield. *J. Prakt. Chem.,* **2000**, *342*(3), 269-280.
[http://dx.doi.org/10.1002/(SICI)1521-3897(200003)342:3<269::AID-PRAC269>3.0.CO;2-0]

[10]   Colacino, E.; Nun, P.; Colacino, F.M.; Martinez, J.; Lamaty, F. Solvent-Free Synthesis of Nitrones in a Ball-Mill. *Tetrahedron,* **2008**, *64*(23), 5569-5573.
[http://dx.doi.org/10.1016/j.tet.2008.03.091]

[11]   Zhu, X.; Li, Z.; Shu, Q.; Zhou, C.; Su, W. Mechanically activated solid state synthesis of flavones by high-speed ball milling. *Synth. Commun.,* **2009**, *39*, 4199-4211.
[http://dx.doi.org/10.1080/00397910902898551]

[12]   Zhu, X.; Li, Z.; Jin, C.; Xu, L.; Wu, Q.; Su, W. Mechanically Activated Synthesis of 1,3,5-triaryl-2-pyrazolines by High Speed Ball Milling. *Green Chem.,* **2009**, *11*(2), 163-165.
[http://dx.doi.org/10.1039/b816788e]

[13]   Zhang, H.; Liu, R-Q.; Liu, K-C.; Li, Q-B.; Li, Q-Y.; Liu, S-Z. A one-pot approach to pyridyl isothiocyanates from amines. *Molecules,* **2014**, *19*(9), 13631-13642.
[http://dx.doi.org/10.3390/molecules190913631] [PMID: 25185069]

[14]   a) Kaupp, G.; Naimi-Jamal, M.R. Quantitative Cascade Condensations between *o*-Phenylenediamines and 1,2-Dicarbonyl Compounds without Production of Wastes. *Eur. J. Org. Chem.,* **2002**, *2002*(8), 1368-1373.
[http://dx.doi.org/10.1002/1099-0690(200204)2002:8<1368::AID-EJOC1368>3.0.CO;2-6] b) Sharma, H.; Singh, N.; Jang, D.O. A Ball-Milling Strategy For The Synthesis Of Benzothiazole, Benzimidazole And Benzoxazole Derivatives Under Solvent-Free Conditions. *Green Chem.,* **2014**, *16*, 4922-4930.
[http://dx.doi.org/10.1039/C4GC01142B]

[15]   a) Oliveira, P.F.; Haruta, N.; Chamayou, A.; Guidetti, B.; Baltas, M.; Tanaka, K.; Sato, T.; Baron, M. Comprehensive Experimental Investigation of Mechanically induced 1, 4-diazines Synthesis in Solid State. *Tetrahedron,* **2017**, *73*, 2305-2310.
[http://dx.doi.org/10.1016/j.tet.2017.03.014] b) Bhutia, Z.T.; Prasannakumar, G.; Das, A.; Biswas, M.; Chatterjee, A.; Banerjee, M.A. Facile, Catalyst-Free Mechano-Synthesis of Quinoxalines and their In-Vitro Antibacterial Activity Study. *ChemistrySelect,* **2017**, *2*, 1183-1187.
[http://dx.doi.org/10.1002/slct.201601672]

[16]   Carlier, L.; Baron, M.; Chamayou, A.; Couarraze, G. Use of Co-grinding as a Solvent-Free Solid State Method to Synthesize Dibenzophenazines. *Tetrahedron Lett.,* **2011**, *52*, 4686-4689.
[http://dx.doi.org/10.1016/j.tetlet.2011.07.003]

[17]   Zeng, J-C.; Xu, H.; Yu, F.; Zhang, Z. Manganese (III) acetate Mediated Synthesis of Polysubstituted Pyrroles under Solvent-Free Ball Milling. *Tetrahedron Lett.,* **2017**, *58*, 674-678.
[http://dx.doi.org/10.1016/j.tetlet.2017.01.016]

[18]   Kaupp, G.; Schmeyers, J.; Boy, J. Quantitative Solid-State Reactions of Amines with Carbonyl Compounds and Isothiocyanates. *Tetrahedron,* **2000**, *56*(36), 6899-6911.
[http://dx.doi.org/10.1016/S0040-4020(00)00511-1]

[19]   Kaupp, G.; Naimi-Jamal, M.R. Kneading Ball-Milling and Stoichiometric Melts for the Quantitative Derivatization of Carbonyl Compounds with Gas–Solid Recovery. *Eur. J. Org. Chem.,* **2002**, *2*(3), 1368.
[http://dx.doi.org/10.1002/1099-0690(200204)2002:8<1368::AID-EJOC1368>3.0.CO;2-6]

[20]   Paveglio, G.C.; Longhi, K.; Moreira, D.N.; München, T.S.; Tier, A.Z.; Gindri, I.M.; Bender, C.R.; Frizzo, C.P.; Zanatta, N.; Bonacorso, H.G. How Mechanical and Chemical Features Affect the Green Synthesis of 1 H-Pyrazoles in a Ball Mill. *ACS Sustain. Chem.& Eng.,* **2014**, *2*, 1895-1901.
[http://dx.doi.org/10.1021/sc5002353]

[21]   Zhang, Z.; Tan, Y-J.; Wang, C-S. One-pot Synthesis of 3,5-diphenyl-1H-pyrazoles from Chalcones and Hydrazine under Mechanochemical Ball Milling. *Heterocycles,* **2014**, *89*, 103-112.
[http://dx.doi.org/10.3987/COM-13-12867]

[22]   Wang, G-W.; Dong, Y-W.; Wu, P.; Yuan, T-T.; Shen, Y-B. Unexpected solvent-free cycloadditions of 1,3-cyclohexanediones to 1-(pyridin-2-yl)-enones mediated by manganese(III) acetate in a ball mill. *J. Org. Chem.,* **2008**, *73*(18), 7088-7095.

[http://dx.doi.org/10.1021/jo800870z] [PMID: 18710288]

[23]   a) Lambat, T.L.; Chaudhary, R.G.; Abdala, A.A.; Mishra, R.K.; Sami, M.; Banerjee, S. Mesoporous PbO Nanoparticles-catalyzed Arylbenzodioxy Xanthenedione Scaffolds under Solvent-less Conditions in a Ball Mill. *RSC Advances,* **2019**, *9*, 31683-31690.
[http://dx.doi.org/10.1039/C9RA05903B] b) Lambat, T.L.; Abdala, A.A.; Sami, M.; Ledad, P.V.; Chaudhary, R.G.; Banerjee, S. Sulfamic acid promoted one-pot multicomponent reaction: a facile synthesis of 4-oxotetrahydroindoles under ball milling conditions. *RSC Advances,* **2019**, *9*, 39735-39742.
[http://dx.doi.org/10.1039/C9RA08478A]

[24]   Thorwirth, R.; Stolle, A.; Ondruschka, B.; Wild, A.; Schubert, U.S. Fast, ligand- and solvent-free copper-catalyzed click reactions in a ball mill. *Chem. Commun. (Camb.),* **2011**, *47*(15), 4370-4372.
[http://dx.doi.org/10.1039/c0cc05657j] [PMID: 21399799]

[25]   Mukherjee, N.; Ahammed, S.; Bhadra, S.; Ranu, B.C. Solvent-Free One Pot Synthesis of 1,2,3-triazole Derivatives by the Click Reaction of Alkyl Halides or Aryl Boronic Acids, Sodium Azide and Terminal Alkynes over a Cu/Al_2O_3 Surface under Ball Milling. *Green Chem.,* **2013**, *15*, 389-397.
[http://dx.doi.org/10.1039/C2GC36521A]

[26]   Štrukil, V.; Margetić, D.; Igrc, M.D.; Eckert-Maksić, M.; Friščić, T. Desymmetrisation of aromatic diamines and synthesis of non-symmetrical thiourea derivatives by click-mechanochemistry. *Chem. Commun. (Camb.),* **2012**, *48*(78), 9705-9707.
[http://dx.doi.org/10.1039/c2cc34013e] [PMID: 22914574]

[27]   Vogel, P.; Figueira, S.; Muthukrishnan, S.; Mack, J. Environmentally benign Nucleophilic Substitution Reactions. *Tetrahedron Lett.,* **2009**, *50*(1), 55-56.
[http://dx.doi.org/10.1016/j.tetlet.2008.10.079]

[28]   Phung, C.; Ulrich, R.M.; Ibrahim, M.; Tighe, N.T.G.; Lieberman, D.L.; Pinhas, A.R. *Green Chem.,* **2011**, *13*, 3224-3229.
[http://dx.doi.org/10.1039/c1gc15850c]

[29]   Thorwirth, R.; Stolle, A. Solvent-Free Synthesis of Enamines from Alkyl Esters of Propiolic or But--yne Dicarboxylic Acid in a Ball Mill. *Synlett,* **2011**, 2200-2202.

[30]   Zille, M.; Stolle, A.; Wild, A.; Schubert, U.S. $ZnBr_2$-Mediated Synthesis of Indoles in a Ball Mill by Intramolecular Hydroamination of 2-alkynylanilines. *RSC Advances,* **2014**, *4*, 13126-13133.
[http://dx.doi.org/10.1039/c4ra00715h]

[31]   Dong, Y-W. Y.-W.; Wang, G.-W.; Wang, L. Solvent-Free Synthesis of Naphthopyrans under Ball-Milling Conditions. *Tetrahedron,* **2008**, *64*(44), 10148-10154.
[http://dx.doi.org/10.1016/j.tet.2008.08.047]

[32]   Waddell, D.C.; Thiel, I.; Bunger, A.; Nkata, D.; Maloney, A.; Clark, T.; Smith, B.; Mack, J. Investigating the Formation of Dialkyl Carbonates using High Speed Ball Milling. *Green Chem.,* **2011**, *13*(11), 3156-3161.
[http://dx.doi.org/10.1039/c1gc15594f]

[33]   Metwally, M. A.; Monatsh. Waste-Free Solid-State Organic Syntheses: Solvent-Free Alkylation, Heterocyclization, and Azo-Coupling Reactions. *Chem,* **2007**, *138*(8), 771-776.

[34]   Gao, J.; Wang, G.W. Direct oxidative amidation of aldehydes with anilines under mechanical milling conditions. *J. Org. Chem.,* **2008**, *73*(7), 2955-2958.
[http://dx.doi.org/10.1021/jo800075t] [PMID: 18331062]

[35]   Chatterjee, T.; Saha, D.; Ranu, B.C. Solvent-Free Transesterification in a Ball-Mill over Alumina Surface. *Tetrahedron Lett.,* **2012**, *53*(32), 4142-4144.
[http://dx.doi.org/10.1016/j.tetlet.2012.05.127]

[36]   Thorwirth, R.; Bernhardt, F.; Stolle, A.; Ondruschka, B.; Asghari, J. Switchable selectivity during oxidation of anilines in a ball mill. *Chemistry,* **2010**, *16*(44), 13236-13242.

[http://dx.doi.org/10.1002/chem.201001702] [PMID: 20922723]

[37]   Szuppa, T.; Stolle, A.; Ondruschka, B.; Hopfe, W. Solvent-free dehydrogenation of γ-terpinene in a ball mill: investigation of reaction parameters. *Green Chem.,* **2010**, *12*, 1288-1294. [http://dx.doi.org/10.1039/c002819c]

[38]   Szuppa, T.; Stolle, A.; Ondruschka, B.; Hopfe, W. An alternative solvent-free synthesis of nopinone under ball-milling conditions: Investigation of reaction parameters. *ChemSusChem,* **2010**, *3*(10), 1181-1191. [http://dx.doi.org/10.1002/cssc.201000122] [PMID: 20737534]

[39]   Mack, J.; Fulmer, D.; Sofel, S.; Santos, N. The first solvent-free method for the reduction of esters. *Green Chem.,* **2007**, *9*, 1041-1043. [http://dx.doi.org/10.1039/b706167f]

CHAPTER 5

Recent Advances in the Developments of Enantioselective Electrophilic Fluorination Reactions *via* Organocatalysis

Kavita Jain and **Kalpataru Das**[*]

Department of Chemistry, School of Chemical Sciences and Technology, Dr. Harisingh Gour University (A Central University), Sagar - 470003 (M.P.), India

Abstract: In recent years, the organocatalytic electrophilic fluorination reactions have been extensively explored for the synthesis of organofluorine compounds. The systematic introduction of fluorine atom often improves a number of properties of fluorinated molecules including metabolic stability and various pharmacological properties and thus frequently employed to design fluorinated drugs. The enantioselective electrophilic fluorination *via* organocatalysis has emerged as the most powerful approach for the synthesis of organofluorine compounds as the organocatalytic approaches have several advantages in terms of economical and environmental benefit. In this chapter, the most important developments of organocatalytic enantioselective electrophilic fluorination are highlighted using new types of electrophilic fluorinating reagents (NFSI, F-TEDA-BF$_4$, F-CA-BF$_4$) in the presence of readily available different types of organocatalysts such as different amine based catalysts, phase-transfer catalysts, Brønsted acid and H-bonding catalysts, which are stable, easy to handle, more efficient and selective. Some recent advances with fascinating examples, mechanism of electrophilic fluorination, mode of activation of catalysts, catalytic cycles, controlling product selectivity and synthesis of chiral fluorine containing drugs have been described.

Keywords: Amine catalysts, Asymmetric fluorination, Biologically active compounds, Catalysis, Electrophilic fluorination, Fluorinating reagents, Green chemistry, Hydrogen bonding catalyst, *N*-fluorobenzenesulfonimide (NFSI), Organocatalyst, Organofluorine compounds, Pharmaceuticals, Phase-transfer catalysis, Selectfluor (F-TEDA-BF$_4$), Quaternary ammonium salts.

INTRODUCTION

The organofluorine compounds have attracted enormous attention to synthetic

[*] **Corresponding author Kalpataru Das:** Department of Chemistry, School of Chemical Sciences and Technology, Dr. Harisingh Gour University (A Central University), Sagar - 470003 (M.P.), India; E-mail: kalpatarud@gmail.com

Goutam Kumar Patra & Santosh Singh Thakur (Eds.)

chemists due to their unique applications in pharmaceuticals [1a], agrochemicals [1b-c] and materials [1d]. The systematic introduction of fluorine atoms into organic molecules has been extensively employed in drug discovery research to improve the pharmacological properties of drug molecules [2]. Accordingly, the selective fluorination of organic compounds has become a fascinating research area nowadays in organic synthesis [3]. Among the fluorination methods, the catalytic electrophilic fluorination is one of the most widely used methodologies for the synthesis of several fluorine containing pharmaceuticals [4]. In fact, 35% of agrochemicals and 20% of pharmaceuticals available on the market contain one or more fluorine atoms in their structures [4c]. In particular, the asymmetric synthesis of α-fluorinated carbonyl compounds with a fluorine atom at a quaternary carbon centre attracted special interest as these compounds exhibit various important biological activities [5]. Some representative biologically active fluorinated compounds are shown in Fig. (**1**).

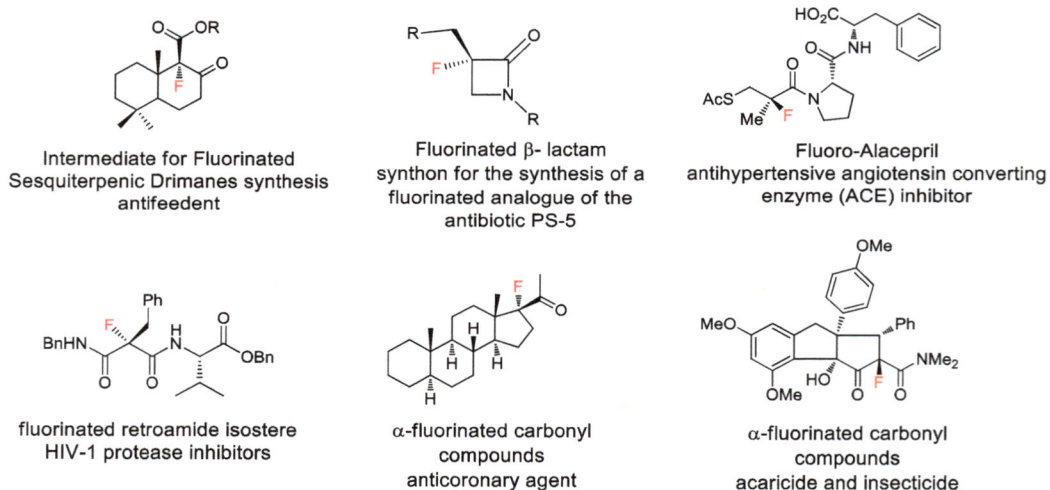

Intermediate for Fluorinated Sesquiterpenic Drimanes synthesis antifeedent

Fluorinated β- lactam synthon for the synthesis of a fluorinated analogue of the antibiotic PS-5

Fluoro-Alacepril antihypertensive angiotensin converting enzyme (ACE) inhibitor

fluorinated retroamide isostere HIV-1 protease inhibitors

α-fluorinated carbonyl compounds anticoronary agent

α-fluorinated carbonyl compounds acaricide and insecticide

Fig. (1). Some representative biologically active fluorinated compounds.

Due to the above facts, in the past decades, enantioselective electrophilic fluorination reactions have been developed effectively both *via* asymmetric organocatalysis and metal catalysis approaches [3]. However, the electrophilic fluorination *via* organocatalysis has emerged as the most powerful approach for the synthesis of organofluorine compounds as the organocatalytic approaches have several advantages in terms of economical and environmental benefits. In this chapter, metal catalysis is not incorporated and only the most important developments on enantioselective electrophilic fluorination of carbonyl and related compounds are highlighted *via* organocatalysis. The electrophilic fluorination reactions were emphasized using new types of electrophilic fluorinating reagents (NFSI, F-TEDA-BF$_4$, F-CA-BF$_{4,}$ *etc.*) in the presence of different types of novel organocatalysts including different amine based catalysts, phase-transfer catalysts, and H-bonding catalysts. Moreover, some recent advances with fascinating examples, mechanism of electrophilic fluorination, mode of activation of catalysts, catalytic cycles, controlling product selectivity and synthesis of chiral fluorine containing drugs have been illustrated.

AN OVERVIEW OF ASYMMETRIC ORGANOCATALYSIS

In recent years, asymmetric organocatalytic transformations have emerged as powerful tools as compared to metal catalyzed reactions due to the high performance of organocatalysts in terms of efficiency and selectivity [6]. Thus, the often asymmetric organocatalytic reaction has been employed in building complex molecular architecture [7]. Typically, the use of organocatalysts has several advantages in terms of economical and environmental benefit as they are tolerant of water and air, and are easily accessible by synthetically as well from the natural sources [6, 8]. Mostly organocatalytic reactions easy to perform without the requirement of any inert atmosphere and special equipment. Due to these advantages, several organocatalytic reactions have been developed in the last decades using a green chemistry approach towards sustainable development in asymmetric synthesis [9]. Almost more than fifty years back it has been revealed that the organocatalyst, O-acetylquinine (1 mol%) derived from quinine as a natural source and used effectively in asymmetric organic transformation and afforded chiral products with good yields and average enantioselectivity (Scheme **1**) [10]. This pioneering work by Pracejus *et al.* has opened a new era in the asymmetric organocatalysis.

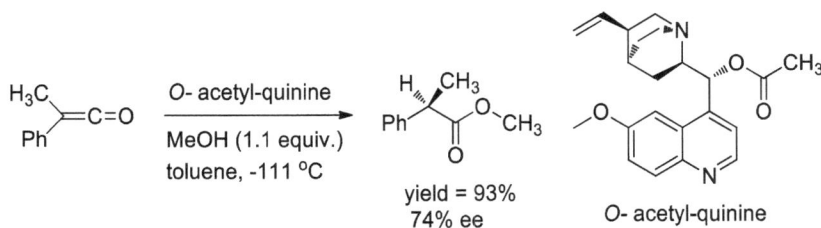

Scheme (1). O-Acetyl quinine catalyzed asymmetric addition of methanol to prochiral ketene.

A number of efficient organocatalysts have been discovered and used in the catalytic and enantioselective synthesis of various functionalized molecules and pharmaceuticals [11 - 17]. Selected examples of organocatalysts are listed in Fig. (**2**), which are used widely in asymmetric synthesis using primary amine based catalysis [11], secondary amine catalysis *via* enamines [12], secondary amine catalysis *via* iminium ions [13], phase transfer catalysis [14a-b], nucleophilic chiral amine catalysis [15], Brønsted acid catalysis [16] and bifunctional H-bonding catalysis [17a].

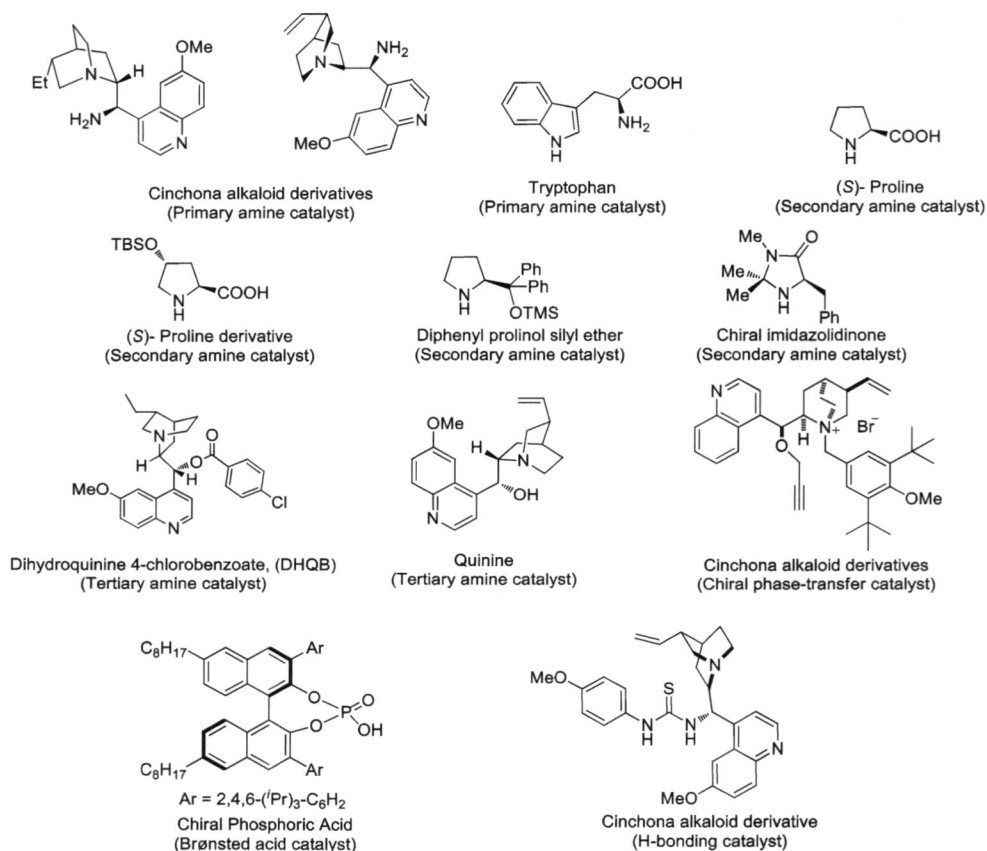

Cinchona alkaloid derivatives
(Primary amine catalyst)

Tryptophan
(Primary amine catalyst)

(S)- Proline
(Secondary amine catalyst)

(S)- Proline derivative
(Secondary amine catalyst)

Diphenyl prolinol silyl ether
(Secondary amine catalyst)

Chiral imidazolidinone
(Secondary amine catalyst)

Dihydroquinine 4-chlorobenzoate, (DHQB)
(Tertiary amine catalyst)

Quinine
(Tertiary amine catalyst)

Cinchona alkaloid derivatives
(Chiral phase-transfer catalyst)

Ar = 2,4,6-(^{i}Pr)$_3$-C$_6$H$_2$
Chiral Phosphoric Acid
(Brønsted acid catalyst)

Cinchona alkaloid derivative
(H-bonding catalyst)

Fig. (2). Selected chiral organocatalysts for enantioselective synthesis.

ORGANOCATALYTIC ENANTIOSELECTIVE ELECTROPHILIC FLUORINATION REACTIONS

Different strategies have been discovered to introduce the fluorine atom into the organic molecules. One of the approaches is the radical-mediated incorporation of fluorine [18], which required very complex reagent system and reaction condition to initiate the radical reaction and sometimes resulting low yields of fluorinated products. Other approach based on nucleophilic fluorination *via* metal catalysis [19], where metal catalysts are very expensive and fluoride anion is less nucleophilic due to high electronegativity of the fluorine atom, and thus less explored. On the other hand, electrophilic fluorination is one of the most attractive strategies for the selective introduction of fluorine atom(s) into organic molecules for synthesis of fluorine containing molecules [3]. In the past decades, various electrophilic fluorinating reagents have been developed for the enantioselective electrophilic fluorination, which are stable, safe and selective reagents as shown in Fig. (**3**) [20].

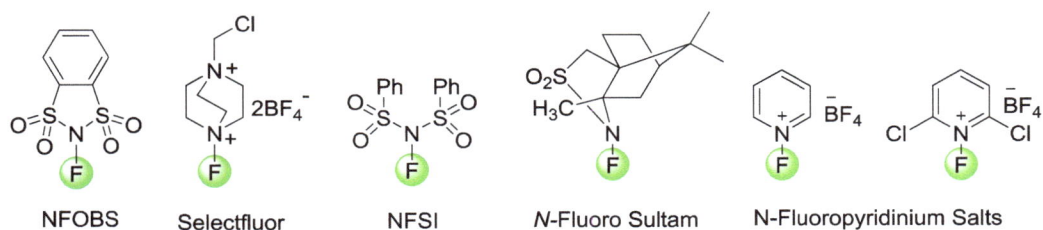

Fig. (3). Electrophilic fluorinating reagents.

Although, a number of fluorination reactions emerged for the asymmetric synthesis of organofluorine compounds using electrophilic fluorination *via* metal catalysis [21], however, such metal catalyst are highly expensive and required special reaction set up as well as precaution to accomplish the reaction. However, electrophilic fluorination using organocatalysts has been appeared as the most powerful and attractive approach for the green and sustainable development in the field of asymmetric synthesis of organofluorine compounds [22]. Due to the limited scope, in this chapter, we emphasized the most important developments on enantioselective electrophilic fluorination of carbonyl and related compounds using electrophilic fluorinating reagents *via* organocatalysis.

Primary Amine Catalyzed Enantioselective Electrophilic Fluorination

The first example of enantioselective α-fluorination of ketones using a cinchona alkaloid based primary amine as organocatalyst has been reported by Macmillan group using electrophilic fluorinating reagent NFSI [23a]. After systematic optimization of the reaction using various amine catalysts, it has been found that Cinchona-based alkaloid catalysts, dihydroquinidine **3** with trichloroacetic acid (TCA) as the cocatalyst provided best results at -10 °C and obtained *R*-fluorocyclohexanone in 88% yield and 99% ee when cyclohexanone was used as substrate (Scheme **2**). Here, enamine activation strategy provides the diastereo-, regio-, and chemoselective catalyst control in α-fluorination of ketones.

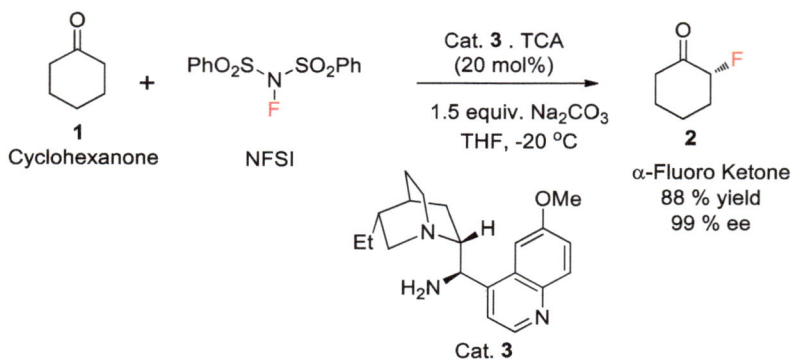

Scheme (2). Primary amine catalyzed enantioselective α-fluorination of cyclic ketone.

This asymmetric α-fluorination strategy successfully employed with various cyclic ketones and afforded fluorinated carbo and hetrocyclic products [23a]. The strategy has also been effectively applied for chemo- and regioslective α-fluorination of polycycle to synthesize complex carbonyl compound α-fluoro-*allo*-pregnanedione **4** with very good yield and stereoselectivity (91% yield, 99% ee) as shown in Scheme (3).

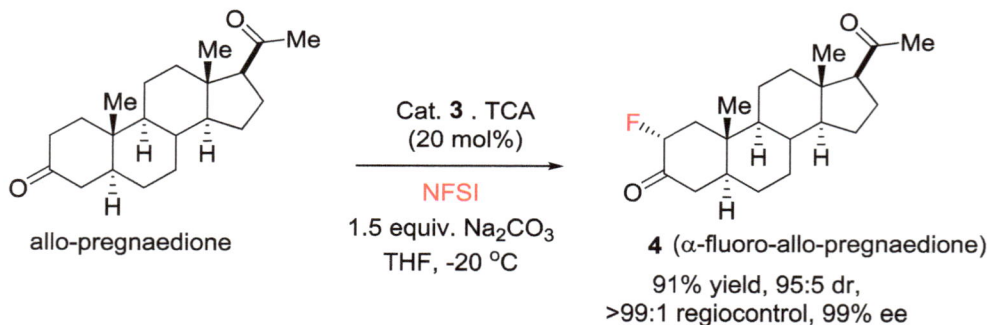

Scheme (3). Chemo- and regioslective α-fluorination polycycle.

Later, Xu and co-workers has been described the enantioselective fluorination of β-ketoesters using chiral primary amine-based multicatalyst system. In this reaction, a cyclic β-ketoester **5** reacted with fluorinating agent Selectfluor at 0 °C in the presence of multicatalyst system to obtain chiral fluorinated α-ketoester **6** as shown in Scheme (4) [23b]. In this studies, it has been revealed that good yield and moderate enantioselectivity (90% yield and 55% ee) of the fluorinated product **6** was obtained when cinchona alkaloid derived catalyst **7** (QN-NH$_2$) in combination with *L*-leucine was used as primary amines based dual organocatalyst.

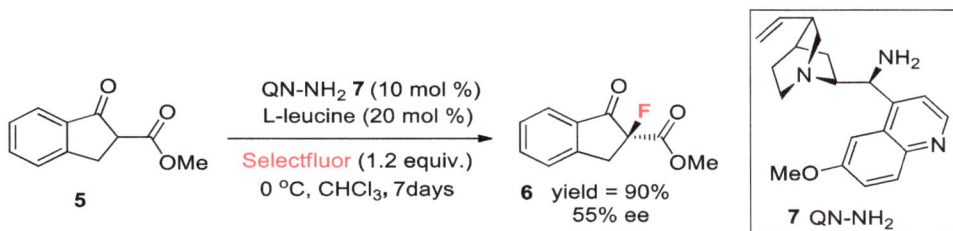

Scheme (4). Enantioselective fluorination of β-ketoester catalyzed by primary amine.

The mechanism of the reaction has been proposed by the formation of enamine and iminium intermediates as shown in Scheme (**5**). When the cinchona alkaloid derived chiral primary amine (QN-NH$_2$) **7** as catalyst reacted with β-ketoester **5**, it generated enamine **8** and iminium ion **9** intermediates. Subsequently, it was reacted with Selectfluor as fluorinating reagent *via* electrophilic fluorination to generate chiral fluorinated β-ketoester **6** [23b].

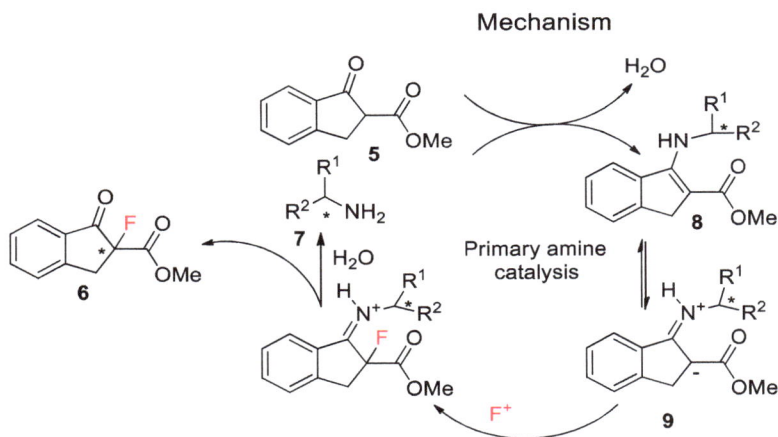

Scheme (5). Possible mechanism of QN-NH$_2$-catalyzed enantioselective fluorination.

Subsequently, Jacobsen *et al.* have been developed enantioselective α-fluorination of α-branched aldehyde **10** using a new primary amine based organocatalyst simple benzamide analog **12** to obtain the fluorinated product **11** in excellent yields and enantioselectivity (Scheme **6**) [24]. Similarly, the α-fluorination of α-branched aldehyde **10** accomplished *via* enamine mechanism. Using the same strategy, however using lower catalyst loading (5 mol%) a gram scale synthesis of fluoroalcohol **11** have been developed with excellent yield and enantioselectivity (>99% yield, 80% ee).

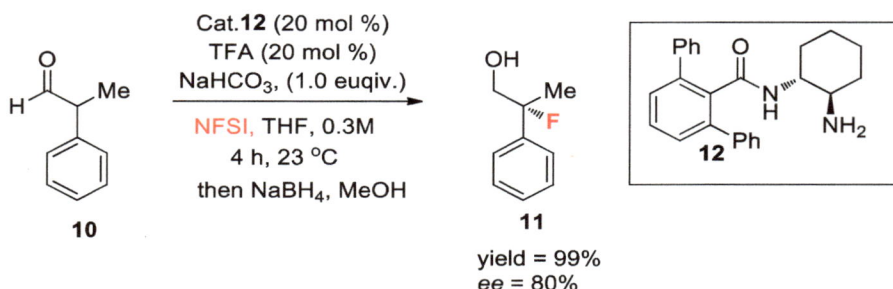

Scheme (6). Enantioselective α-fluorination of α-branched aldehyde

In 2016, Shibatomi *et al.* reported a highly enantioselective fluorination of α-branched aldehyde **13** using new class of binaphthyl based chiral primary amine as organocatalyst **15** to afford the corresponding fluoroaldehyde, which was easily converted into chiral fluorinated primary alcohol **14** with good yield 86% and enantioselectivity 95% (Scheme **7**). The reaction proceeded *via* enantioselective electrophilic fluorination with *N*-fluorobenzenesulfonimide (NFSI) as fluorinating agent in the presence of 10 mol% of catalyst **12** and 10 mol% 3,5- dinitrobenzoic acid as a co-catalyst [25].

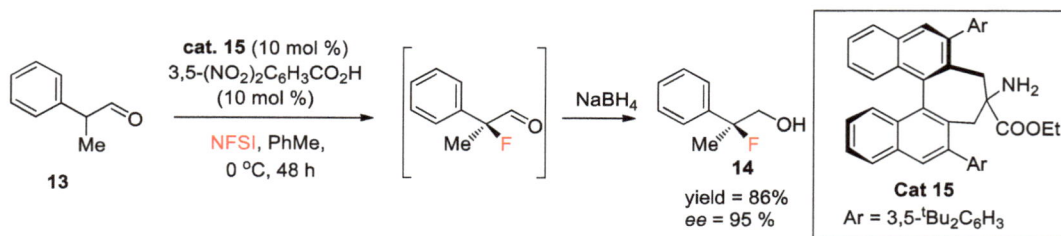

Scheme (7). Chiral primary amine catalyzed α-fluorination of α-branched aldehyde.

Recently, Luo *et al.* developed the reagent-controlled asymmetric α-fluorination of β-keto ester **16** and it was revealed that asymmetric fluorination effectively controls the enantioselectivity of fluorinated product by simple varying the fluorinating reagent in the presence of chiral primary amine catalyst **17**. In this strategy, both enantiomers with fluorine atom at quaternary carbon centre, **18** and **19** were formed with good yield and enantioselectivity *via* asymmetric electrophilic fluorination using NFSI and *N*-fluoro-pyridinium salts, respectively, as shown in Scheme (**8**). The mechanism for the product formation has been proposed through dual H-bonding and electrostatic stereocontrolling modes for a single chiral primary amine catalyst [26].

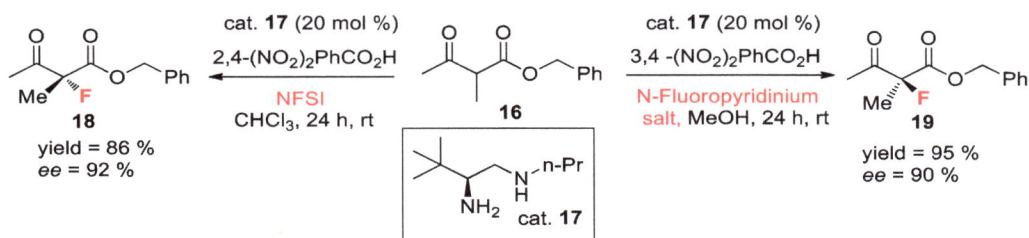

Scheme (8). Chiral primary amine catalyzed α-fluorination reactions of *β*-keto ester.

Secondary Amine Catalyzed Enantioselective Electrophilic Fluorination

Secondary amines based enamine catalysis represents an important area of organocatalysis and has been exploited extensively for the synthesis of chiral organofluorine compounds. Among the secondary amine based catalysts proline and its derivative have shown promising catalytic activity for α-fluorination of carbonyl compounds to generate chiral fluorinated molecules with quaternary stereogenic centre in an enantiocontrolled way. In 2005, Enders and coworker reported the first organocatalytic direct asymmetric α-fluorination of aldehyde and ketones using Selectfluor as electrophilic fluorinating agent in the presence of (*S*)-proline and proline derivatives as secondary amine based organocatalyst to obtain the corresponding chiral α-fluorocarbonyls in moderate to good yields (up to 78% yield), however, the enantioslectivity of the α-fluorination reaction remained low (up to 36% ee) [27a]. In the same year, subsequently, MacMillan *et al.* have reported the direct ennatioselective α-fluorination of aldehyde *via* enamine catalysis. A range of aldehydes **20** underwent ennatioselective α-fluorination using commercially available NFSI as electrophilic fluorinating agent and chiral imidazolidinone-dichloroacetic acid (DCA) salt **21** as the secondary amine orgaocatalyst. In this protocol, the unstable fluorinated aldehydes were directly transformed into various chiral fluorinated primary alcohols **22** using NaBH$_4$ without loss of any enantioselectivity as shown in Scheme (**9**) [27b].

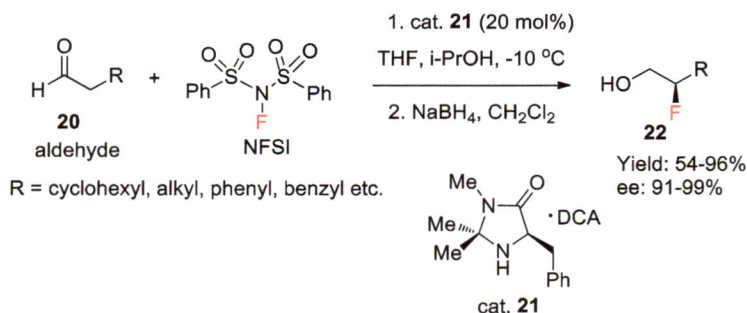

Scheme (9). Enantioslective α-fluorination of aldehydes.

Concurrently in 2005, Jørgensen and co-workers reported the enantioselective α-fluorination of aldehydes **23** using organocatalytic electrophilic C-F bond forming reaction *via* enamine catalysis. The reaction was effectively catalyzed by silylated prolinol derivative **24** (1 mol%) as secondary amine based organocatalyst using NFSI as fluorine source in methyl tert-butyl ether (MTBE) as solvent. In this strategy a range of aldehydes underwent enantioselective α-fluorination reaction; however, the product was isolated as α-fluoroalcohols **26** with good yields and enantioselectivities (91-97% ee) as shown in Scheme (**10**) [27c].

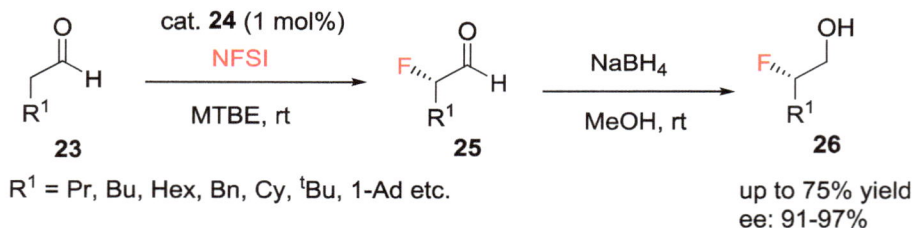

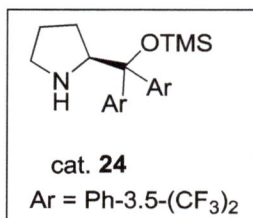

Scheme (10). Organocatalytic enantioslective α-fluorination of aldehydes using catalyst **24**.

The intial products α-fluoroaldehydes **25** as shown in Scheme (**10**) were reduced in situ directly to **26** without loss of enantioselectivity. It has been observed that α-fluoroaldehydes **25** are unstable as they decomposed easily in silica gel column and are also volatile in nature.

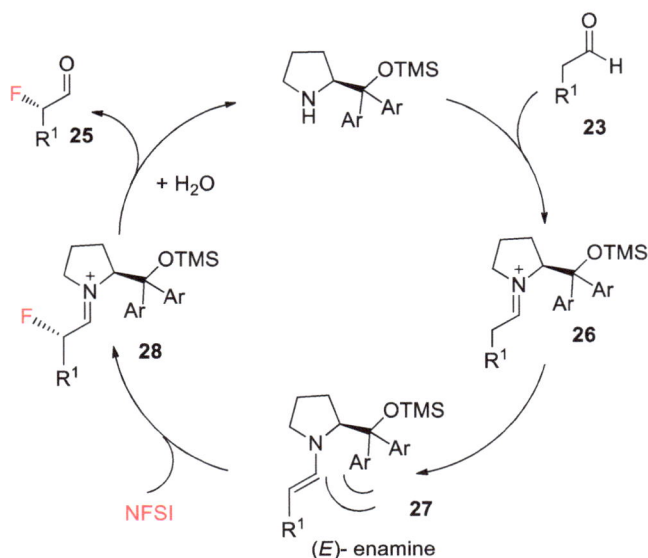

Scheme (11). Proposed organocatalytic cycle for α-fluorination of aldehydes.

The authors described the mechanism of α-fluorination of aldehydes *via* enamine as represented in Scheme (**11**) to rationalize the stereochemical outcome. They proposed that the formation of *E*-isomer of enamine **27** as the more stable intermediate, where it has been showed that one of the 3,5-di(trifluoromethyl)phenyl groups covers the *Re* face of the enamine. As a consequence, the electrophilic C-F bond formation takes place from *Si* face of **27**, and provided excellent enantioselectivities (up to 97%). Moreover, they extended the scope of the reaction and modified the protocol for α-fluorination of sterically crowded branched aldehydes. The sterically crowded aldehyde **29** required less bulky catalyst **30** (5 mol%) and higher temperature to facilitate the electrophilic C-F bond formation and generated chiral α-fluoroalcohol **32** in 78% yield and 48% ee in the same way (Scheme **12**) [27c].

Similarly, in 2005 Barbas and coworker have reported α-fluorination of branched and linear aldehydes for the synthesis of chiral quaternary α-fluoroaldehyde in high yield and moderate enantioselectivity *via* enamine catalysis using imidazolidinones as chiral promoters [27d].

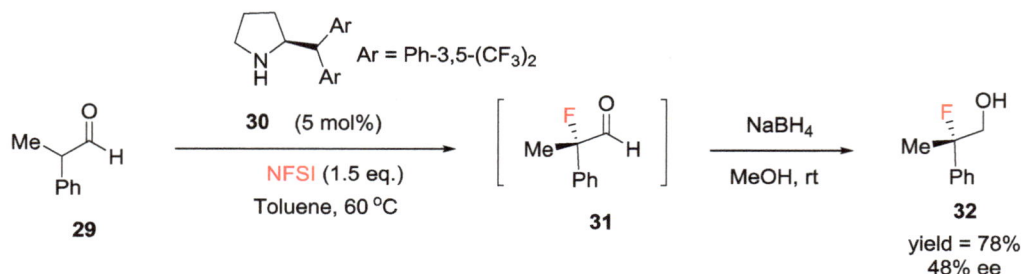

Scheme (12). Organocatalytic α-fluorination of strerically crowded aldehydes.

Later in 2009, Jørgensen and co-workers have reported a simple, direct one-pot organocatalytic approach for the synthesis of optically active allylic and propargylic fluorides (Scheme **13**) [27e]. This method is based on organocatalytic α-fluorination of aldehyde **33** with NFSI in the presence of the proline based secondary amine chiral organocatalyst **34** (1 mol%) as the catalyst. Subsequently, α-fluoroaldehyde formed in the reaction has been trapped *in-situ* through homologation using Ohira-Bestmann reagent **35** in an one-pot reaction condition as shown in Scheme **13**. The generalization of the fluorination reaction is accomplished by the formation of a range of nonracemic propargylic fluorides **36** in good yields and excellent enantioselectivities. The optically active propargylic fluorides is a key intermediate for the synthesis of monofluorinated leukotrienes **37** as shown in Scheme (**13**) [27e].

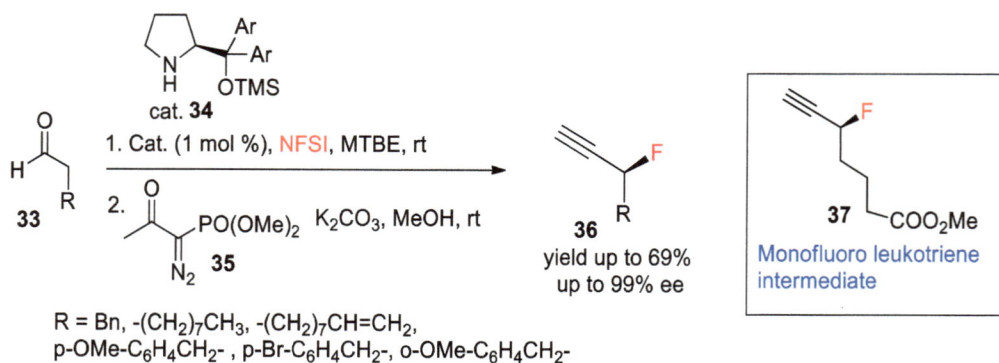

Scheme (13). One-pot synthesis of optically active propargylic fluorides.

The authors proposed the mechanism for the formation of propargylic fluorides **36** as shown in Scheme (**14**). Catalytic reaction initiated with the condensation of the catalyst **34** and aldehyde **33** leading to the formation of a reactive iminium ion **38** and enamine species **39**, which favors electrophilic fluorination from the *Si*-face due to steric shielding of the *Re*-face by the catalyst. The fluorinated aldehyde **41** is trapped by the Ohira-Bestmann reagent which leads to formation of intermediate **42** and **43**. Finally, the propargylic fluoride **36** is formed by a 1,2-proton shift and elimination of N$_2$ from intermediate **43** [27e].

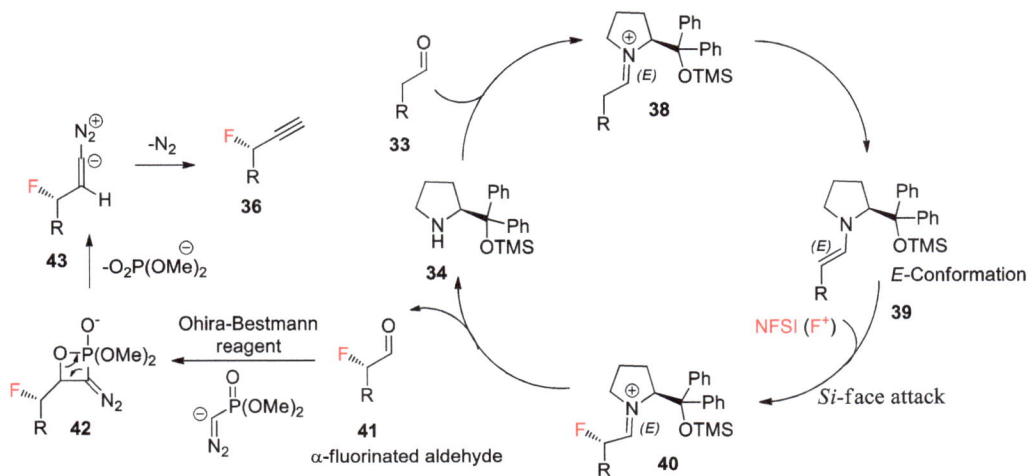

Scheme (14). Mechanism of the formation of optically active propargylic fluorides.

Furthermore, the strategy has been extended for the synthesis of nonracemic allylic fluorides. It was also reported that nonracemic fluorinated triazoles were obtained from aldehyde **33** using click-chemistry in one-pot procedures [27e].

Tertiary Amine Catalyzed Enantioselective Electrophilic Fluorination

The chiral fluorinating reagents have been developed long back for reagent controlled asymmetric synthesis of organofluoro compounds, and these are mostly relied on using stoichiometric amount of chiral *N*-fluoro reagents in electrophilic fluorination [28]. In 2000, Shibata and coworkers [29a] and Cahard and coworkers [29b] have reported their research works independently and disclosed a fundamentally new class of chiral fluorinating reagents, which are derived from naturally occurring cinchona alkaloids. These chiral *N*-fluoroammonium salts (F-CA-BF$_4$) are stable and isolable and can be easily derived through combination of cinchona alkaloids (CAs) with Selectfluor. These chiral *N*-fluoro reagents are very efficient for the enantioselective fluorination of enol ethers like silyl enol ethers and metal enolates. Nevertheless, these fluorination reactions are not catalytic and required stoichiometric amounts of cinchona alkaloids as tertiary amine based organocatalyst [29]. These pioneering research works, however, provided new directions for the development of catalytic enantioselective electrophilic fluorination using cinchona alkaloids based tertiary amine based organocatalyst. Consequently, in 2006, Shibata and co-workers reported for the first time a novel catalytic approach for the enantioselective fluorination of acyl enol ethers **44** using chiral *N*-fluoroammonium salts derived from the combination of cinchona alkaloid derivative dihydroquinine 4-chlorobenzoate (DHQB) **46** and Selectfluor for the synthesis of non-racemic α-fluoroketones **45** in good yields and enantioselectivity as shown in Scheme (**15**) [30].

Scheme (15). Enantioselective electrophilic fluorination of acyl enol ethers.

The catalytic cycle for the enantioselective electrfluorination reaction is shown in Scheme (**16**). It was found that acyl enol ethers **44** are not reactive enough towards Selectfluor under the reaction condition, which allowed the generation of reactive *N*-fluoroammonium salts **47** as chiral fluorinating agent by the combination between cinchona alkaloid DHQB **46** and Selectfluor.

Scheme (16). Catalytic cycle for the enantioselective fluorination of acyl enol ethers.

The activation of enol ether **44** was required by using NaOAc as base to effect enatioselective fluorination of enol ether **44** with reactive chiral *N*-fluoro reagents **47** and followed by trapping of the acetyl cation (Ac$^+$) and BF$_4^-$ in the reaction cycle to regenerate the catalyst **46** [30]. Using the methodology described above [29a], the first enantioselective synthesis of a potent potassium channel opener BMS-204352 **49** (MaxiPost) was reported by Shibata *et al.* using cinchona alkaloid **50**/Selectfluor mediated enantiselective fluorination of oxindole **48**. Enantioselective fluorination provided **49** with very good yield and enantioselectivity (84% ee) as shown in Scheme (**17**) [31].

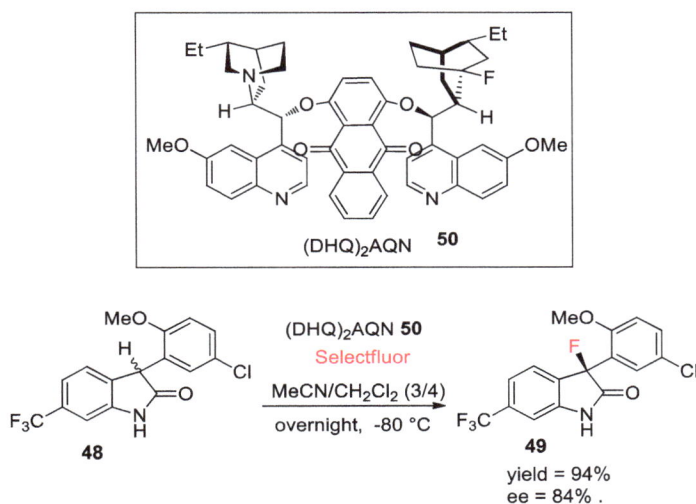

Scheme (17). Enantioselective Synthesis of BMS-204352 using chiral *N*-fluoroammonium salt.

The asymmetric fluorination of 4-substituted pyrazolones **51** catalyzed by quinine **53** as a tertiary amine based organocatalysts was revealed in the presence of *N*-fluorobenzenesulfonimide as fluorinating agent afforded various 4-fluoropyrazo--5-ones **52** with very good yield and enantioselectivity as shown in Scheme (**18**) [32]. The 4-pyrazol-5-ones is one of the main scaffold found in HIV-1 integrase inhibitor **54** (Scheme **18**).

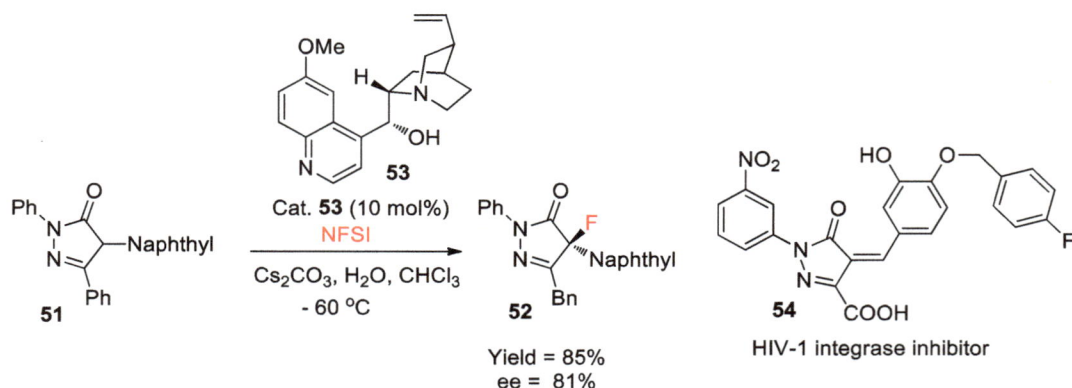

Scheme (18). Asymmetric fluorination of 4-substituted pyrazolones catalyzed by quinine.

Enantioselective Electrophilic Fluorination *via* Phase Transfer Catalysis (PTC)

Chiral quaternary ammonium salts are known as chiral phase-transfer catalysts and has been proved as versatile catalyst systems used widely in organic synthesis and synthesis of pharmaceutical products [14]. Due to several advantages, asymmetric phase-transfer catalysis (PTC) become much popular in green and sustainable chemistry as various asymmetric bond forming reactions have been achieved using PTC under mild reaction conditions [33]. Two representative reaction systems is considered for phase-transfer-catalyzed asymmetric C-F bond formations using natural cinchonidine based chiral quaternary ammonium salts as chiral phase-transfer catalysts as illustrated in Schemes (**19** and **20**). In 2002, Kim *et al.* reported the first organocatalytic enantioselective fluorination of cyclic *β*-keto ester **55** using NFSI as electrophilic fluorinating agent and **56** (10 mol%) as chiral phase transfer catalyst in toluene with mild base K$_2$CO$_3$ at room temperature [34].

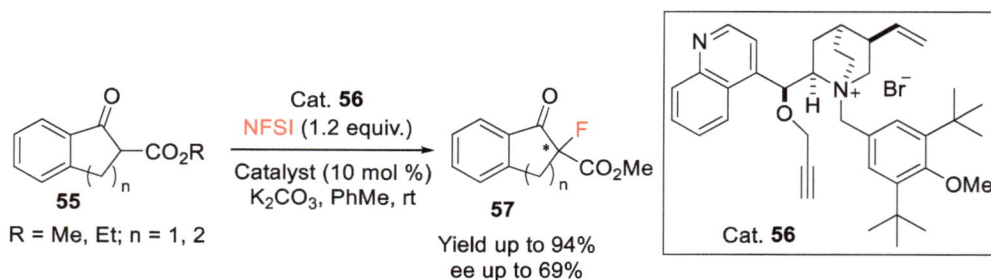

Scheme (19). Enantioselective fluorination of β-ketoester under phase transfer catalysis.

It was found that selection of large group at the quinuclidine nitrogen of cinchona alkaloid was vital for high stereoselectivity. Under phase-transfer catalysis, various cyclic β-ketoester afforded fluorinated cyclic α-fluoro-β-ketoesters **57** with high yields and good to moderate enantioselectivity up to 69% ee as shown in Scheme (**19**) [34]. Subsequently, in 2004 Kim group developed catalytic enantioselective electrophilic fluorination of α-cyanoacetate derivative **58** with *N*-fluorobenzenesulfonimide using the same catalyst **56** (10 mol%) under mild phase transfer conditions. The aryl substituted α-cyano acetate derivative **58** was reacted *via* asymmetric electrophilic C-F bond forming reaction to give the corresponding chiral α-cyano-α-fluoro acetate derivative **59** in high yield with good enantiomeric excess (73% ee) as shown in Scheme (**20**) [35].

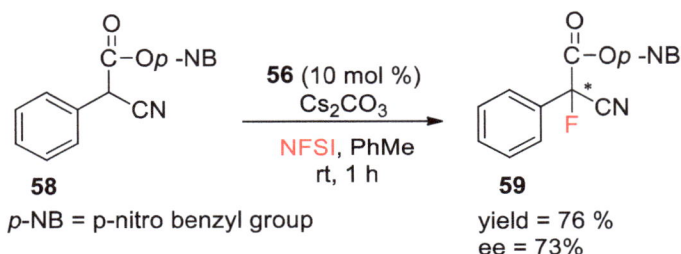

Scheme (20). Enantioselective fluorination of α-cyano ester under phase transfer catalysis.

Later, Lu and co-workers, reported that the 1-adamantoyl derivative of cinchona alkaloid **60** as chiral phase transfer catalyst for the efficient enantioselective electrophilic fluorination of cyclic β-keto ester **61**. When *t*-butyl ester group was present on indanone carboxylate **67** under biphasic reaction condition in aqueous K_2CO_3 and toluene as solvent at -25 °C afforded the chiral α-fluorinated β-keto ester **62** with good yield and enantioselectivity (90% ee) in the presence **60** (10 mol%) using NFSI as a fluorinating reagent (Scheme **21**) [36].

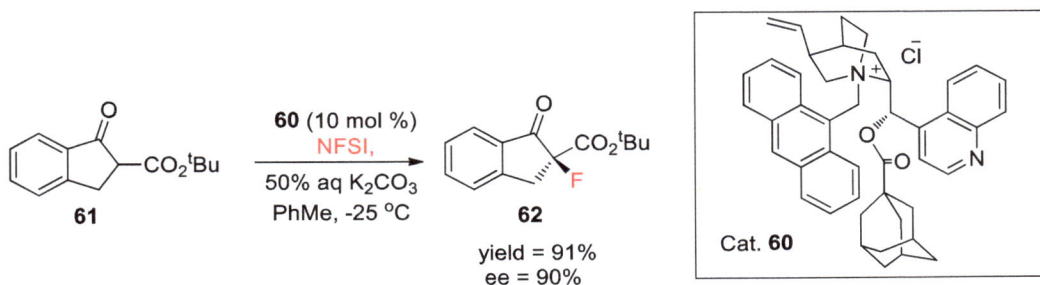

Scheme (21). Enantioselective fluorination of β- ketoester under phase transfer catalysis.

In 2014, Toste and co-worker reported enantioselective electrophilic fluorination of α-branched cyclohexanone **63** using Selectfluor as fluorinating agent to obtain fluorinated product **64** with quaternary fluorine-containing stereocentres *via* chiral anion phase transfer catalysis in combination with enamine catalysis as shown in Schemes (**22** to **23**) [37]. In this asymmetric transformation, the fluorination is achieved through the combination of two different chiral organocatalytic cycles with enamine activation of ketone using chiral protected amino acid **A** as organocatalyst and activation of Selectfluor by chiral phosphoric acid (CPA), a Brønsted acid, which acts as chiral anion phase transfer catalyst **65**. Subsequently, two different chiral activated components are combined and interacted directly *via* H-bonding and forming a transition state which is responsible for observed enantioselectivity (Scheme **23**). The strategy has been successfully applied to various carbo and heterocyclic α-branched cyclohexanone to give the desired fluorinated products **64** with high ee up to 94% ee [37].

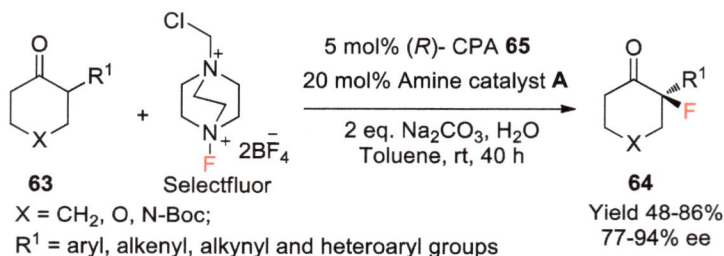

Scheme (22). Asymmetric fluorination of α-branched cyclic ketone.

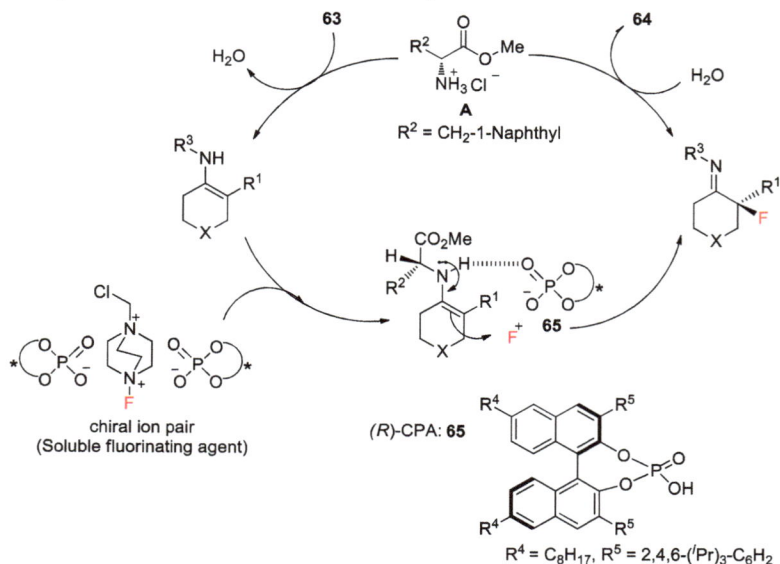

Scheme (23). Catalytic cycle for asymmetric fluorination *via* chiral anion phase transfer catalysis.

Enantioselective Electrophilic Fluorination *via* Hydrogen Bonding Catalysis

The use of cinchona alkaloid based organocatalysis has been revealed as efficient approach as these organocatalysts often offering high stereoselective transformations under mild reaction conditions. In 2012, Niu *et al.* reported that the cinchona alkaloid based bifunctional thiourea catalyst **66** is highly effective for enantioselective fluorination of β-ketoester **67** (Scheme **24**).

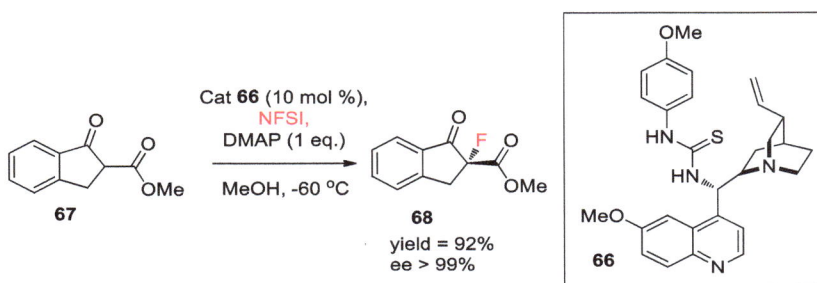

Scheme (24). Thiourea-catalyzed enantioselective fluorination of cyclic β-keto ester.

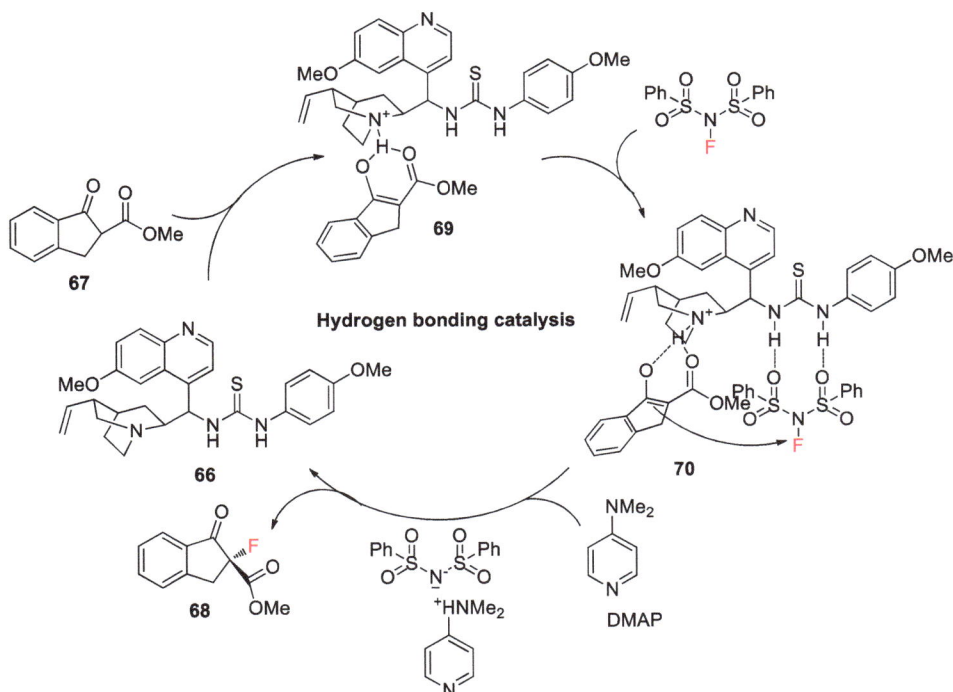

Scheme (25). Catalytic cycle for enantioselective fluorination *via* hydrogen bonding catalysis.

Authors found the more than 99% enantiomeric excess of fluorinated β-ketoester **68** when they used 10 mol % catalyst **66** in methanol at -60 °C with 4-Dimethylaminopyridine (DMAP) as an additive and NFSI as a fluorinating reagent. They also observed the substituent effects on enantioselectvity and observed above 90% *ee*, when alkoxy groups of indanonecarboxylate was methyl or benzyl group. Surprisingly, when Selectfluor was used instead of NFSI, they have obtained racemic product only and concluded that the obtained results through proposed mechanism as shown in Scheme (**25**), where fluorination was proceedings *via* dual activation by the bifunctional thiourea catalyst **66** with substrate β-ketoester **67** as well as with NFSI through hydrogen bonding interactions as shown in the intermediate **70** (Scheme **25**) [38].

CONCLUDING REMARKS

In the last decades, remarkable advances of organocatalysis have been made with the manifestation of various efficient organocatalytic and enantioselective fluorination reactions. The tremendous powers of organocatalysis have been revealed through synthesizing range of nonracemic fluorinated organic compounds. A wide range of organocatalytic enantioselctive electrophilic fluorination of organic compounds have been demonstrated using new type of commercially available fluorinating reagents in the presence of readily available organocatalysts such as different amine based catalysts, phase-transfer catalyst, Brønsted acid and H-bonding catalysts. These organocatalytic electrophilic fluorination processes provided useful fluorinated products including chiral active methylene compounds, carbonyl compounds having fluorine atom at quaternary carbon centre and fluorinated streogenic centre. Several fascinating examples with mechanism of electrophilic fluorination, mode of activation by the catalysts, catalytic cycles have been represented. It was highlighted that organocatalytic approaches have several advantages and are attractive in the field of green and sustainable asymmetric synthesis. With the development of new strategy for incorporation fluorine, a few examples with practical routes of complex fluorinated targets having biological importance have been included.

CONSENT FOR PUBLICATION

Not applicable.

CONFLICT OF INTEREST

The author declares that there is no conflict of interest in this chapter.

ACKNOWLEDGEMENTS

The authors are grateful to the University Grants Commission (UGC), New Delhi for Start-up grant (No. F.30-56/2014 BSR) and Science and Engineering Research Board (SERB), Govt. of India for research grant (EMR/2017/000234) for generous financial support. We thank the Department of Chemistry, Dr. Harisingh Gour University, Sagar (M.P.) for providing infrastructure facilities. KJ is thankful to UGC for the award of MANF research fellowship.

REFERENCES

[1]   a) O'Hagan, D. Fluorine in health care: Organofluorine containing blockbuster drugs. *J. Fluor. Chem.,* **2010**, *131*, 1071.
[http://dx.doi.org/10.1016/j.jfluchem.2010.03.003] b) Jeschke, P. Latest generation of halogen-containing pesticides. *Pest Manag. Sci.,* **2017**, *73*(6), 1053-1066.
[http://dx.doi.org/10.1002/ps.4540] [PMID: 28145087] c) Fujiwara, T.; O'Hagan, D. Successful fluorine-containing herbicide agrochemicals. *J. Fluor. Chem.,* **2014**, *167*, 16.
[http://dx.doi.org/10.1016/j.jfluchem.2014.06.014] d) Liang, Y.; Feng, D.; Wu, Y.; Tsai, S.T.; Li, G.; Ray, C.; Yu, L. Highly efficient solar cell polymers developed *via* fine-tuning of structural and electronic properties. *J. Am. Chem. Soc.,* **2009**, *131*(22), 7792-7799.
[http://dx.doi.org/10.1021/ja901545q] [PMID: 19453105]

[2]   a) Mei, H.; Han, J.; Klika, K.D.; Izawa, K.; Sato, T.; Meanwell, N.A.; Soloshonok, V.A. Applications of fluorine-containing amino acids for drug design. *Eur. J. Med. Chem.,* **2020**, *186*111826
[http://dx.doi.org/10.1016/j.ejmech.2019.111826] [PMID: 31740056] b) Yerien, D.E.; Bonesi, S.; Postigo, A. Fluorination ethods in drug discovery. *Org. Biomol. Chem.* **2016**, *14*, 8398. (b) Hagmann, W. K. *J. Med. Chem.,* **2008**, *51*, 4359.c) Kirk, K.L. The Many Roles for Fluorine in Medicinal Chemistry. *Org. Process Res. Dev.,* **2008**, *12*, 305.
[http://dx.doi.org/10.1021/op700134j]

[3]   a) Moschner, J.; Stulberg, V.; Fernandes, R.; Huhmann, S.; Leppkes, J.; Koksch, B. Approaches to Obtaining Fluorinated Amino Acids. *Chem. Rev.,* **2019**, *119*(18), 10718-10801.
[http://dx.doi.org/10.1021/acs.chemrev.9b00024] [PMID: 31436087] b) Yang, X.; Wu, T.; Phipps, R.J.; Toste, F.D. Advances in catalytic enantioselective fluorination, mono-, di-, and trifluoromethylation, and trifluoromethylthiolation reactions. *Chem. Rev.,* **2015**, *115*(2), 826-870.
[http://dx.doi.org/10.1021/cr500277b] [PMID: 25337896] c) Liang, T.; Neumann, C.N.; Ritter, T. Introduction of Fluorine and Fluorine□Containing Functional Groups. *Angew. Chem. Int. Ed.,* **2013**, *52*, 8214.
[http://dx.doi.org/10.1002/anie.201206566] d) Lectard, S.; Hamashima, Y.; Sodeoka, M. Fluorination in Medicinal Chemistry: Methods, Strategies, and Recent Developments. *Adv. Synth. Catal.,* **2010**, *352*, 2708.
[http://dx.doi.org/10.1002/adsc.201000624] e) Valero, G.; Companyo, X.; Rios, R. Enantioselective organocatalytic synthesis of fluorinated molecules. *Chemistry,* **2011**, *17*(7), 2018-2037.
[http://dx.doi.org/10.1002/chem.201001546] [PMID: 21294172] f) Bobbio, C.; Gouverneur, V. Catalytic asymmetric fluorinations. *Org. Biomol. Chem.,* **2006**, *4*(11), 2065-2075.
[http://dx.doi.org/10.1039/b603163c] [PMID: 16729117] g) Lin, J-H.; Xiao, J-C. Recent advances in asymmetric fluorination and fluoroalkylation reactions *via* organocatalysis. *Tetrahedron Lett.,* **2014**, *55*, 6147-6155.
[http://dx.doi.org/10.1016/j.tetlet.2014.09.031]

[4]   a) Zhu, Y.; Han, J.; Wang, J.; Shibata, N.; Sodeoka, M.; Soloshonok, V.A.; Coelho, J.A.S.; Toste, F.D. Modern Approaches for Asymmetric Construction of Carbon-Fluorine Quaternary Stereogenic Centers: Synthetic Challenges and Pharmaceutical Needs. *Chem. Rev.,* **2018**, *118*(7), 3887-3964.
[http://dx.doi.org/10.1021/acs.chemrev.7b00778] [PMID: 29608052] bZhou, Y.; Wang, J.; Gu, Z.; Wang, S.; Zhu, W.; Aceña, J. L.; Soloshonok, V. A.; Izawa, K.; Liu, H. *Chem. Rev.,* **2016**, *116*, 422.c)

Wang, J.; Sánchez-Roselló, M.; Aceña, J.L.; del Pozo, C.; Sorochinsky, A.E.; Fustero, S.; Soloshonok, V.A.; Liu, H.; Liu, H. Fluorine in pharmaceutical industry: fluorine-containing drugs introduced to the market in the last decade (2001-2011). *Chem. Rev.,* **2014**, *114*(4), 2432-2506.
[http://dx.doi.org/10.1021/cr4002879] [PMID: 24299176]

[5]   a) Shibata, N.; Ishimaru, T.; Suzuki, E.; Kirk, K.L. Enantioselective fluorination mediated by N-fluoroammonium salts of cinchona alkaloids: first enantioselective synthesis of BMS-204352 (MaxiPost). *J. Org. Chem.,* **2003**, *68*(6), 2494-2497.
[http://dx.doi.org/10.1021/jo026792s] [PMID: 12636425] b) Li, J. Cai, Y.; Chem, W.; Liu, X.; Lin, L.; Feng, X. *J. Org. Chem.,* **2012**, *77*, 9148.
[http://dx.doi.org/10.1021/jo301705t] [PMID: 23030737] c) Phan, L.T.; Clark, R.F.; Rupp, M.; Or, Y.S.; Chu, D.T.W.; Ma, Z. Synthesis of 2-fluoro-6-O-propargyl-11,12-carbamate ketolides. A novel class of antibiotics. *Org. Lett.,* **2000**, *2*(19), 2951-2954.
[http://dx.doi.org/10.1021/ol006226o] [PMID: 10986080] d) Shibata, N.; Ishimaru, T.; Nakamura, M.; Toru, T. 20-Deoxy-20-fluorocamptothecin: Design and Synthesis of Camptothecin Isostere. *Synlett,* **2004**, *14*, 2509.
[http://dx.doi.org/10.1055/s-2004-834810]

[6]   Berkessel, A. Groeger, H. *Asymmetric Organocatalysis: from Biomimetic Concepts to Applications in Asymmetric Synthesis*; Wiley-VCH, **2005**. b) Dalko, P.I.; Moisan, L. In the Golden Age of Organocatalysis. *Angew. Chem. Int. Ed.,* **2004**, *43*, 5138-5175.
[http://dx.doi.org/10.1002/anie.200400650] c) Seayad, J.; List, B. Asymmetric organocatalysis. *Org. Biomol. Chem.,* **2005**, *3*(5), 719-724.
[http://dx.doi.org/10.1039/b415217b] [PMID: 15731852] d) Dondoni, A.; Massi, A. Asymmetric Organocatalysis: From Infancy to Adolescence. *Angew. Chem. Int. Ed.,* **2008**, *47*, 4638.
[http://dx.doi.org/10.1002/anie.200704684]

[7]   a) Ishikawa, H.; Shiomi, S. Alkaloid synthesis using chiral secondary amine organocatalysts. *Org. Biomol. Chem.,* **2016**, *14*(2), 409-424.
[http://dx.doi.org/10.1039/C5OB02021B] [PMID: 26625722] b) Pellissier, H. Recent Developments in Asymmetric Organocatalytic Domino Reaction. *Adv. Synth. Catal.,* **2012**, *354*, 237.
[http://dx.doi.org/10.1002/adsc.201100714] c) Borissov, A.; Davies, T.Q.; Ellis, S.R.; Fleming, T.A.; Richardson, M.S.W.; Dixon, D.J. Organocatalytic enantioselective desymmetrisation. *Chem. Soc. Rev.,* **2016**, *45*(20), 5474-5540.
[http://dx.doi.org/10.1039/C5CS00015G] [PMID: 27347567] d) Gaunt, M.J.; Johansson, C.C.; McNally, A.; Vo, N.T. Enantioselective organocatalysis. *Drug Discov. Today,* **2007**, *12*(1-2), 8-27.
[http://dx.doi.org/10.1016/j.drudis.2006.11.004] [PMID: 17198969] e) Marqués-López, E.; Herrera, R.P.; Christmann, M. Asymmetric organocatalysis in total synthesis--a trial by fire. *Nat. Prod. Rep.,* **2010**, *27*(8), 1138-1167.
[http://dx.doi.org/10.1039/b924964h] [PMID: 20445939]

[8]   a) Xu, L.W.; Lu, Y. Primary amino acids: privileged catalysts in enantioselective organocatalysis. *Org. Biomol. Chem.,* **2008**, *6*(12), 2047-2053.
[http://dx.doi.org/10.1039/b803116a] [PMID: 18528563] b) Marcelli, T.; Hiemstra, H. Cinchona Alkaloids in Asymmetric Organocatalysis. *Synthesis,* **2010**, 1229.
[http://dx.doi.org/10.1055/s-0029-1218699] c) Tian, S.K.; Chen, Y.; Hang, J.; Tang, L.; McDaid, P.; Deng, L. Asymmetric organic catalysis with modified cinchona alkaloids. *Acc. Chem. Res.,* **2004**, *37*(8), 621-631.
[http://dx.doi.org/10.1021/ar030048s] [PMID: 15311961]

[9]   a) Hernández, J.G.; Juaristi, E. Recent efforts directed to the development of more sustainable asymmetric organocatalysis. *Chem. Commun. (Camb.),* **2012**, *48*(44), 5396-5409.
[http://dx.doi.org/10.1039/c2cc30951c] [PMID: 22517403] b) Raj, M.; Singh, V.K. Organocatalytic reactions in water. *Chem. Commun. (Camb.),* **2009**, *44*(44), 6687-6703.
[http://dx.doi.org/10.1039/b910861k] [PMID: 19885454]

[10]   a) Pracejus, H. Organische Katalysatoren, LXI. Asymmetrische Synthesen mit Ketenen, I. Alkaloid□katalysierte asymmetrische Synthesen von α□Phenyl□propionsäureestern. *Justus Liebigs*

Ann. Chem., **1960**, *634*, 9.
[http://dx.doi.org/10.1002/jlac.19606340103] b) Pracejus, H.; Mätje, H. Organische Katalysatoren. LXXI Asymmetrische Synthesen mit Ketenen. IV. Zusammenhänge zwischen dem räumlichen Bau einiger alkaloidartiger Katalysatoren und ihren stereospezifischen Wirkungen bei asymmetrischen Estersynthesen . *J. Prakt. Chem.,* **1964**, *24*, 195.
[http://dx.doi.org/10.1002/prac.19640240311]

[11]   a) Xu, L.W.; Luo, J.; Lu, Y. Asymmetric catalysis with chiral primary amine-based organocatalysts. *Chem. Commun. (Camb.),* **2009**, (14), 1807-1821.
[http://dx.doi.org/10.1039/b821070e] [PMID: 19319412] b) Duan, J.; Li, P. Asymmetric organocatalysis mediated by primary amines derived from cinchona alkaloids: recent advances. *Catal. Sci. Technol.,* **2014**, *4*, 311.
[http://dx.doi.org/10.1039/C3CY00739A]

[12]   a) Mukherjee, S.; Yang, J.W.; Hoffmann, S.; List, B. Asymmetric enamine catalysis. *Chem. Rev.,* **2007**, *107*(12), 5471-5569.
[http://dx.doi.org/10.1021/cr0684016] [PMID: 18072803] b) Bertelsen, S.; Jørgensen, K.A. Organocatalysis--after the gold rush. *Chem. Soc. Rev.,* **2009**, *38*(8), 2178-2189.
[http://dx.doi.org/10.1039/b903816g] [PMID: 19623342]

[13]   Erkkilä, A.; Majander, I.; Pihko, P.M. Iminium catalysis. *Chem. Rev.,* **2007**, *107*(12), 5416-5470.
[http://dx.doi.org/10.1021/cr068388p] [PMID: 18072802]

[14]   a) Ooi, T.; Maruoka, K. Recent Advances in Asymmetric Phase□Transfer Catalysis. *Angew. Chem. Int. Ed.,* **2007**, *46*, 4222.
[http://dx.doi.org/10.1002/anie.200601737] b) Hashimoto, T.; Maruoka, K. Recent development and application of chiral phase-transfer catalysts. *Chem. Rev.,* **2007**, *107*(12), 5656-5682.
[http://dx.doi.org/10.1021/cr068368n] [PMID: 18072806] c) Jain, K.; Das, K. A convenient method for the synthesis of fluorinated α-cyanoacetates via phase-transfer catalysis. *Synth. Commun.,* **2018**, *48*, 1966.
[http://dx.doi.org/10.1080/00397911.2018.1473442]

[15]   France, S.; Guerin, D.J.; Miller, S.J.; Lectka, T. Nucleophilic chiral amines as catalysts in asymmetric synthesis. *Chem. Rev.,* **2003**, *103*(8), 2985-3012.
[http://dx.doi.org/10.1021/cr020061a] [PMID: 12914489]

[16]   Akiyama, T.; Mori, K. Stronger Brønsted Acids: Recent Progress. *Chem. Rev.,* **2015**, *115*(17), 9277-9306.
[http://dx.doi.org/10.1021/acs.chemrev.5b00041] [PMID: 26182163]

[17]   a) Doyle, A.G.; Jacobsen, E.N. Small-molecule H-bond donors in asymmetric catalysis. *Chem. Rev.,* **2007**, *107*(12), 5713-5743.
[http://dx.doi.org/10.1021/cr068373r] [PMID: 18072808] b) Jain, K.; Chaudhuri, S.; Pal, K.; Das, K. The Knoevenagel condensation using quinine as an organocatalyst under solvent-free conditions. *New J. Chem.,* **2019**, *43*, 1299.
[http://dx.doi.org/10.1039/C8NJ04219E]

[18]   Yan, H.; Zhu, C. Recent advances in radical-mediated fluorination through C–H and C–C bond cleavage. *Sci. China Chem.,* **2017**, *60*, 214.
[http://dx.doi.org/10.1007/s11426-016-0399-5]

[19]   Hollingworth, C.; Gouverneur, V. Transition metal catalysis and nucleophilic fluorination. *Chem. Commun. (Camb.),* **2012**, *48*(24), 2929-2942.
[http://dx.doi.org/10.1039/c2cc16158c] [PMID: 22334339]

[20]   Li, M.; Zheng, H.; Xue, X.S.; Cheng, J.P. Ordering the relative power of electrophilic fluorinating, trifluoromethylating, and trifluoromethylthiolating reagents: A summary of recent efforts. *Tetrahedron Lett.,* **2018**, *59*, 1278.
[http://dx.doi.org/10.1016/j.tetlet.2018.02.039]

[21]   a) Hinterman, L.; Togni, A. Catalytic Enantioselective Fluorination of β□Ketoesters. *Angew. Chem.*

Int. Ed., **2000**, *39*, 4359.
[http://dx.doi.org/10.1002/1521-3773(20001201)39:23<4359::AID-ANIE4359>3.0.CO;2-P] b)
Hamashima, Y.; Yagi, K.; Takano, H.; Tamas, L.; Sodeoka, M. An efficient enantioselective
fluorination of various beta-ketoesters catalyzed by chiral palladium complexes. *J. Am. Chem. Soc.,*
2002, *124*(49), 14530-14531.
[http://dx.doi.org/10.1021/ja028464f] [PMID: 12465951] c) Ma, J-A.; Cahard, D. Copper(II) triflate-
bis(oxazoline)-catalysed enantioselective electrophilic fluorination of β-ketoesters. *Tetrahedron
Asymmetry,* **2004**, *15*, 1007.
[http://dx.doi.org/10.1016/j.tetasy.2004.01.014] d) Surya Prakash, G.K.; Beier, P. Construction of
Asymmetric Fluorinated Carbon Centers. *Angew. Chem. Int. Ed.,* **2006**, *45*, 2172.
[http://dx.doi.org/10.1002/anie.200503783] e) Brunet, V.A.; Hagan, D.O. Catalytic Asymmetric
Fluorination Comes of Age. *Angew. Chem. Int. Ed.,* **2008**, *47*, 1179.
[http://dx.doi.org/10.1002/anie.200704700] f) Hollingworth, C.; Hazari, A.; Hopkinson, M.N.;
Tredwell, M.; Benedetto, E.; Huiban, M.; Gee, A.D.; Brown, J.M. Gouverneur. V.
Palladium☐Catalyzed Allylic Fluorination. *Angew. Chem. Int. Ed.,* **2011**, *50*, 2613.
[http://dx.doi.org/10.1002/anie.201007307]

[22]   a) Lin, J-H.; Xiao, J-C. Recent advances in asymmetric fluorination and fluoroalkylation reactions via
organocatalysis. *Tetrahedron Lett.,* **2014**, *55*, 6147.
[http://dx.doi.org/10.1016/j.tetlet.2014.09.031] b) Ye, Z.; Zhao, G. Asymmetric synthesis of fluorine-
containing compounds using organocatalysts. *Chimia (Aarau),* **2011**, *65*(12), 902-908.
[http://dx.doi.org/10.2533/chimia.2011.902] [PMID: 22273370]

[23]   a) Kwiatkowski, P.; Beeson, T.D.; Conrad, J.C.; MacMillan, D.W.C. Enantioselective organocatalytic
α-fluorination of cyclic ketones. *J. Am. Chem. Soc.,* **2011**, *133*(6), 1738-1741.
[http://dx.doi.org/10.1021/ja111163u] [PMID: 21247133] b) Shang, J-Y.; Li, L.; Lu, Y.; Yang, K-F.;
Xu, L-W. Enantioselective fluorination of α-branched aldehydes and subsequent conversion to α-
hydroxyacetals via stereospecific C–F bond cleavage. *Synth. Commun.,* **2014**, *44*, 101.
[http://dx.doi.org/10.1080/00397911.2013.791697]

[24]   Witten, M.R.; Jacobsen, E.N. A Simple Primary Amine Catalyst for Enantioselective Hydroxylations
and Fluorinations of Branched Aldehydes. *Org. Lett.,* **2015**, *17*(11), 2772-2775.
[http://dx.doi.org/10.1021/acs.orglett.5b01193] [PMID: 25952578]

[25]   Shibatomi, K.; Kitahara, K.; Okimi, T.; Abe, Y.; Iwasa, S. Enantioselective fluorination of α-branched
aldehydes and subsequent conversion to α-hydroxyacetals via stereospecific C–F bond cleavage.
Chem. Sci. (Camb.), **2016**, *7*, 1388.
[http://dx.doi.org/10.1039/C5SC03486H]

[26]   Yang'en, Y.; Zhang, L.; Luo, S. Reagent-controlled enantioselectivity switch for the asymmetric
fluorination of β-ketocarbonyls by chiral primary amine catalysis. *Chem. Sci. (Camb.),* **2017**, *8*, 621.
[http://dx.doi.org/10.1039/C6SC03109A]

[27]   a) Enders, D.; Hüttl, M.R.M. Direct Organocatalytic α-Fluorination of Aldehydes and Ketones.
Synlett, **2005**, 991.
[http://dx.doi.org/10.1055/s-2005-864813] b) Beeson, T.D.; Macmillan, D.W. Enantioselective
organocatalytic alpha-fluorination of aldehydes. *J. Am. Chem. Soc.,* **2005**, *127*(24), 8826-8828.
[http://dx.doi.org/10.1021/ja051805f] [PMID: 15954790] c) Marigo, M.; Fielenbach, D.; Braunton, A.;
Kjærsgaard, A.; Jørgensen, K.A. Enantioselective Formation of Stereogenic Carbon–Fluorine Centers
by a Simple Catalytic Method. *Angew. Chem. Int. Ed.,* **2005**, *44*, 3703.
[http://dx.doi.org/10.1002/anie.200500395] d) Steiner, D.D.; Mase, N.; Barbas, C.F., III Direct
Asymmetric α☐Fluorination of Aldehydes. *Angew. Chem. Int. Ed.,* **2005**, *44*, 3706.
[http://dx.doi.org/10.1002/anie.200500571] e) Jiang, H.; Falcicchio, A.; Jensen, K.L.; Paixão, M.W.;
Bertelsen, S.; Jørgensen, K.A. Target-directed organocatalysis: a direct asymmetric catalytic approach
to chiral propargylic and allylic fluorides. *J. Am. Chem. Soc.,* **2009**, *131*(20), 7153-7157.
[http://dx.doi.org/10.1021/ja901459z] [PMID: 19419172]

[28]   a) Davis, F.A.; Zhou, P.; Murphy, C.K. Asymmetric fluorination of enolates with N-fluoro 2,10- (3,3-
dichlorocamphorsultam). *Tetrahedron Lett.,* **1993**, *34*, 3971.

[http://dx.doi.org/10.1016/S0040-4039(00)60592-0] b) Differding, E.; Lang, R.W. New fluorinating reagents - I. The first enantioselective fluorination reaction. *Tetrahedron Lett.,* **1988**, *29*, 6087.
[http://dx.doi.org/10.1016/S0040-4039(00)82271-6]

[29]   a) Shibata, N.; Suzuki, E.; Takeuchi, Y. A Fundamentally New Approach to Enantioselective Fluorination Based on Cinchona Alkaloid Derivatives/Selectfluor Combination. *J. Am. Chem. Soc.,* **2000**, *122*, 10728.
[http://dx.doi.org/10.1021/ja002732x] b) Cahard, D.; Audouard, C.; Plaquevent, J-C.; Roques, N. Design, synthesis, and evaluation of a novel class of enantioselective electrophilic fluorinating agents: N-fluoro ammonium salts of cinchona alkaloids (F-CA-BF(4)). *Org. Lett.,* **2000**, *2*(23), 3699-3701.
[http://dx.doi.org/10.1021/ol006610l] [PMID: 11073679]

[30]   Fukuzumi, T.; Shibata, N.; Sugiura, M.; Nakamura, S.; Toru, T. Enantioselective fluorination mediated by cinchona alkaloids/selectfluor combinations: A catalytic approach. *J. Fluor. Chem.,* **2006**, *127*, 548.
[http://dx.doi.org/10.1016/j.jfluchem.2006.01.004]

[31]   Shibata, N.; Ishimaru, T.; Suzuki, E.; Kirk, K.L. Enantioselective fluorination mediated by N-fluoroammonium salts of cinchona alkaloids: first enantioselective synthesis of BMS-204352 (MaxiPost). *J. Org. Chem.,* **2003**, *68*(6), 2494-2497.
[http://dx.doi.org/10.1021/jo026792s] [PMID: 12636425]

[32]   Bao, X.; Wei, S.; Zou, L.; Song, Y.; Qu, J.; Wang, B. Asymmetric fluorination of 4-substituted pyrazolones catalyzed by quinine. *Tetrahedron Asymmetry,* **2016**, *27*, 436.
[http://dx.doi.org/10.1016/j.tetasy.2016.03.013]

[33]   a) Liu, S.; Kumatabara, Y.; Shirakawa, S. Chiral quaternary phosphonium salts as phase-transfer catalysts for environmentally benign asymmetric transformations. *Green Chem.,* **2016**, *18*, 331.
[http://dx.doi.org/10.1039/C5GC02692J] b) Maruoka, K. Designer chiral phase-transfer catalysts for green sustainable chemistry. *Pure Appl. Chem.,* **2012**, *84*, 1575.
[http://dx.doi.org/10.1351/PAC-CON-11-09-31]

[34]   Kim, D.Y.; Park, E.J. Catalytic enantioselective fluorination of beta-keto esters by phase-transfer catalysis using chiral quaternary ammonium salts. *Org. Lett.,* **2002**, *4*(4), 545-547.
[http://dx.doi.org/10.1021/ol010281v] [PMID: 11843587]

[35]   Park, E.J.; Kim, H.R.; Joung, C.U.; Kim, D.Y. Catalytic Enantioselective Fluorination of α-Cyano Esters by Phase-Transfer Catalysis Using Chiral Quaternary Ammonium Salts. *Bull. Korean Chem. Soc.,* **2004**, *25*, 1451.
[http://dx.doi.org/10.5012/bkcs.2004.25.10.1451]

[36]   Luo, J.; Wu, W.; Xu, L-W.; Meng, Y.; Lu, Y. Enantioselective direct fluorination and chlorination of cyclic β-ketoesters mediated by phase-transfer catalysts. *Tetrahedron Lett.,* **2013**, *54*, 2623.
[http://dx.doi.org/10.1016/j.tetlet.2013.03.028]

[37]   Yang, X.; Phipps, R.J.; Toste, F.D. Asymmetric fluorination of α-branched cyclohexanones enabled by a combination of chiral anion phase-transfer catalysis and enamine catalysis using protected amino acids. *J. Am. Chem. Soc.,* **2014**, *136*(14), 5225-5228.
[http://dx.doi.org/10.1021/ja500882x] [PMID: 24684209]

[38]   Xu, J.; Hu, Y.; Huang, D.; Wang, K-H.; Xu, C.; Niu, T. Thiourea☐Catalyzed Enantioselective Fluorination of β☐Keto Esters. *Adv. Synth. Catal.,* **2012**, *254*, 515.
[http://dx.doi.org/10.1002/adsc.201100660]

CHAPTER 6

A Green and Sustainable Biocatalytic Routes to Prepare Biobased Polyols as a Precursor For Polyurethanes as Compared to Existing Biobased Polyol Technology

Bhaskar Sharma[*], **Hema Tandon**, **Pathik Maji** and **Arti Shrivastava**

Department of Chemistry, Guru Ghasidas Vishwavidyalaya, Bilaspur (Chhattsgarh) 495009, India

Abstract: There has been great interest in the replacement of petroleum-based polyols with biobased polyols in polyurethane applications. However, current products mainly triglyceride-based polyols have many drawbacks, and also do not have the structural efficiency and physical performance characteristics that restrict them to limited applications.

To minimize all these limitations, the synthesis of low molecular weight liquid polyols can be performed *via* a robust, simple, environmentally friendly, and solvent-free 'biocatalytic route'. In contrast to chemical methods, enzyme-catalyzed reactions proceed with high enantio- and regioselectivity, under mild conditions, avoiding protection deprotection steps, providing an attractive alternative to conventional chemical methods. Bio-renewable, non-toxic and low-cost monomers, such as 2,5 dihydroxymethyl furan, 1,4 butanediol, glycerol, diglycerol, isosorbide, D-mannitol, D-sorbitol, citric acid, maleic acid, fumaric acid, succinic acid, glutaric acid, adipic acid, sebacic acid and many more can be employed in preparing biobased polyol prepolymers *via* enzymatic catalysis.

Keywords: Biobased polyols, Enzyme catalysis, Polyurethanes.

INTRODUCTION

Polyurethanes have gained too much attention, and they have been referred to as "Jack of all trades" because of their tremendous use in many industries like automotive, furniture, constructions, footwear, and thermal insulations, *etc*. In general, polyurethanes are obtained by the reaction of polyisocyanates with

---

[*] **Corresponding author Bhaskar Sharma:** Department of Chemistry Guru Ghasidas Vshwavidyalaya (A Central University) Koni, Bilaspur (Chhattsgarh) 495009, India; Tel: +91-7999182918; E-mail: bsharma05@gmail.com.

Goutam Kumar Patra & Santosh Singh Thakur (Eds.)

polyols [1]. The conventional polyols used in polyurethanes are obtained from petroleum feedstocks. The global market for polyurethanes is USD 65.5 billion in 2018 [2] and the corresponding market for polyols was estimated at USD 23.6 billion in 2017and it is expected to grow about 8.5% over the period [3].

The economic concerns and strict environmental regulations have driven polyurethane industries to find new bio-renewable, non-toxic, environmentally-friendly, solvent-free, low cost, and efficient polyol's technology.

Also, polyurethane formulations are prepared generally with a high proportion (approx. 40-60% by weight) of organic solvents such as DMF (N, N-dimethyl formamide), MEK (methyl ethyl ketone), acetone, toluene, and para-xylene as the main ingredient. Consequently, such preparation of polyurethanes emits volatile organic compounds (VOC) in the atmosphere and causing pollution and severe health problems [4, 5].

There have been strong interests in the development of bio-based Polyol technology. Among them, triglyceride based polyol technology has been given much preference. This technology can be used in polyurethane applications and requires less energy and may have lower market prices compared to petro-based polyols.

Companies that have developed and marketed hydroxylated vegetable oils as polyol components in polyurethanes include Cargill, Dow Chemical, Bayer Material Science, Urethane Soy Systems, and Vertellus Performance Materials. These companies have used natural oils from soybean, corn, canola, sunflower, and linseed oil. Then, chemical methods were used to introduce the desired hydroxyl content.

Conventional chemical methods employed to introduce hydroxyl groups into vegetable oils include: (i) epoxidation of triglycerides at unsaturated sites followed by ring-opening to obtain polyols with secondary hydroxyl groups [6, 7], (ii) hydroformylation of triglycerides to add aldehyde groups to unsaturated bonds followed by hydrogenation to prepare polyols with primary hydroxyl groups [8, 9] (iii) ozonolysis of triglycerides to produce polyols with terminal primary hydroxyl groups [10, 11].

Thus, current commercial biobased polyols have one or more of the following drawbacks: i) limiting the basic building block to triglycerides restricts the rigidity, degree of hydroxylation, and molecular weight of polyols; ii) triglyceride modification is performed by chemical catalysis often requiring solvents and toxic chemicals at elevated temperatures; iii) varying content of un-reacted saturated fatty acid present in most of the triglycerides results in dangling after the reaction with isocyanates and leads to poor polyurethane properties.

There has been a strong interest in the replacement of petroleum-based polyols with biobased polyols in polyurethane applications. However, current products mainly triglyceride-based polyols have many drawbacks, as discussed above, and also don't have the structural efficiency and physical performance characteristics that restrict them to limited applications.

To minimize all these limitations, the synthesis of low molecular weight liquid polyols is proposed *via* a robust, simple, environmentally friendly, and solvent-free 'biocatalytic route'. In contrast to chemical methods, enzyme-catalyzed reactions proceed with high enantio- and regioselectivity, under mild conditions, avoiding protection deprotection steps, providing an attractive alternative to conventional chemical methods. Moreover, efforts are made to develop solvent-free or solvent-less polyurethane processes because of growing concern on compliance with governmental and environmental regulations that limit the use of volatile organic compounds (VOC). The emphasis has to be given to prepare liquid polyols with sufficient fluidity for solvent-free formulation as well as hydroxyl content and M_n values that are suitable for attaining desired physicomechanical properties of polyurethanes.

BACKGROUND

There is an increasing awareness globally that how to halt the imbalanced flow of carbon contributing to air pollution, global warming, smog formation, ozone depletion, *etc*. The environmental issues and economic competitiveness have forced entire industries to find new, benign, and alternate chemicals derived from renewable resources.

Current Biobased Polyol Technology

Epoxidation of vegetable oils is normally performed using either peroxyacetic or peroxyformic acid [12]. These peracids are prepared by *in situ* reactions between hydrogen peroxide and either acetic or formic acid in the presence of an acid catalyst. Resulting epoxidized natural oils are ring-opened by reactions with methanol or water catalyzed by acid catalysts (organic or inorganic) or by hydrogenation. Natural-oil derived polyol prepolymers obtained by ring-opening with alcohols are liquid, while those prepared by ring-opening with HCl, HBr, or hydrogenation are solid at room temperature [13]. An example in which this technology was commercialized is Cargill's soy-based BiOH™ polyols where ring-opening of epoxides with methanol led to liquid polyol prepolymers used in formulations to prepare flexible foams, coatings, adhesives, and elastomers (Scheme **1**) [14]. This technology won the 2007 presidential green chemistry challenge award. However, the inability of this technology to allow further variation and fine-tuning of polyol polyester structure have forced polyurethane manufacturers to mix 10 to 30% BiOH™ with 90 to 70% of petroleum-derived polyol prepolymers to attain desired polyurethane product performance. This shows a gap in current technology that will be addressed in this chapter [15].

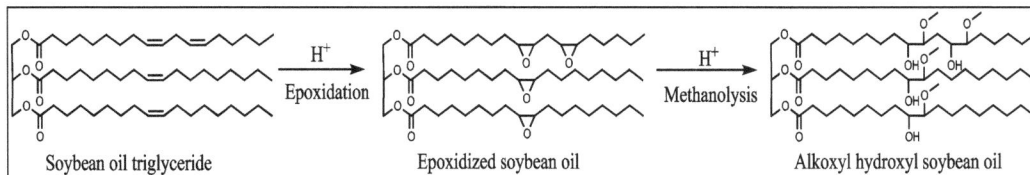

Scheme (1). Cargill's soy-based BiOH™ polyol technology involves epoxidation and subsequent methanolysis to introduce hydroxyl functionality into soybean oil triglycerides [13].

An alternative route developed to hydroxylate triglycerides is by hydroformylation [16a-e] (See Scheme **2**). A fundamental difference in the outcome between hydroformylation and epoxide ring-opening is that, resulting hydroxyl groups are primary and secondary, respectively.

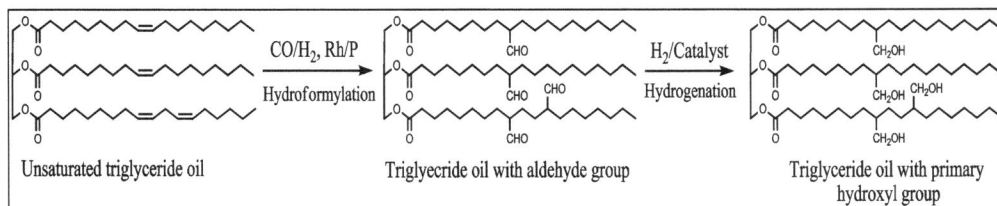

Scheme (2). Polyol-prepolymer preparation by hydroformylation of Vegetable Oils [13].

Primary hydroxyl groups are more reactive than those with secondary substitution which substantially changes process parameters during polyurethane synthesis. Hydroformylation involves: *i*) introduce aldehydes at triglyceride double bonds by the reaction of carbon monoxide and hydrogen (syngas) catalyzed by rhodium or cobalt catalysts at 70-130 °C at high pressures (4 000-11 000 kPa or 40-110 bar) and *ii*) converting aldehyde to hydroxyl groups by hydrogenation. An example of a commercial biobased polyol-prepolymer manufactured by this methodology is RENUVA™, introduced by the Dow Chemical Company. Similar to Cargill, RENUVA™ biobased polyols are targeted for use in polyurethane formulations to make flexible foams in furniture, bedding, automotive, and carpeting as well as polyurethane-based coatings, adhesives, sealants and elastomers [17]. Although rhodium catalysts are very efficient and give high conversions to hydroxylated triglycerides, they are expensive. Indeed, Cobalt catalysts are less expensive than those from rhodium but have much lower catalytic activity leading to triglycerides with lower hydroxyl number than that theoretically expected (*e.g.* ~160 mg KOH/g *versus* 250 mgKOH/g) [13]. Other problems associated with cobalt catalyzed hydroformylation reactions include double bond migration along fatty acid chains leading to less defined and predictable product structures/properties and the occurrence of transesterification reactions [13].

Metathesis has also been explored as a route to further diversify structures of triglyceride derived polyol-prepolymers. The metathesis of triolein with ethylene, catalyzed by a Grubbs ruthenium catalyst, produced a modified triglyceride with terminal carbon-carbon double bonds and 1-decene as a byproduct [18]. The modified triglyceride was epoxidized and epoxy groups were ring-opened with methanol to give terminal hydroxyl groups (Scheme 3).

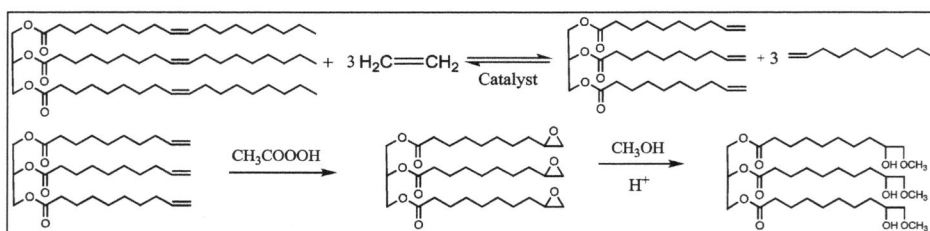

Scheme (3). Polyol-prepolymer preparation by Metathesis of triolein (a model of a triglyceride of oleic acid).

The resulting polyols have higher hydroxyl numbers, lower equivalent weights, lower viscosity, and lower molecular weight values than triglyceride derived polyol prepolymers prepared *via* epoxidation-methanolysis or hydroformylation-hydrogenation routes [19]. While this enables the preparation of polyurethanes without dangling chains, the preparation of polyols by this method is tedious and expensive and, therefore, not commercially viable. This is similarly true for polyol-prepolymers with terminal primary hydroxyl groups prepared by ozonolysis. Vegetable oil ozonolysis is normally performed in methylene chloride/methanol solvent mixtures at -30 to -40 °C [20]. Ozone with oxygen as a carrier gas is passed into reactions to convert triglycerides to triglyceride aldehydes which are then reduced using sodium borohydride. Unlike conventional ozonolysis as depicted above, Tran *et. al* [21] developed an improved method by which ozone was passed through a soybean oil-ethylene glycol solution in the presence of an alkaline catalyst to give ethylene glycol ester primary hydroxyl groups for subsequent reactions with isocyanates.

Other reported routes to polyol prepolymers from triglycerides involve modification by transesterification or transamidation reactions forming lower molecular weight products with higher hydroxyl contents. Typically, transesterifications were conducted between hydroxylated agro-derived triglycerides and an alditol such as glycerol or sorbitol. The high hydroxyl contents of these products target their use in polyurethane formulations to produce rigid insulation foams [22a-b]. Also, increased reactivity of secondary hydroxyl groups within hydroxylated soybean oil triglycerides was achieved by their reaction with ethylene oxide (EO) using acid catalysts (HBF_4, BF_3, CF_3SO_3H, HPF_6, etc) at 35-45 °C at 0.1-0.2 MPa [23]. Furthermore, Luo et. al [24] prepared natural oil derived polyols by aminolysis of triglyceride ester bonds with hydroxyl alkyl amines. Aminolysis reactions were performed by chemical catalysis at elevated temperatures using organometallic catalysts to prepare amine-containing polyol-prepolymers with high hydroxyl numbers (240-530 mg KOH/g) and intermediate viscosities (260-5300 cP at 25 °C). When used in polyurethane formulations, the resulting polyurethanes were said to show improved hardness.

Status of The Existing Bio-Based Technology

One of the most important targets of chemical industries worldwide is the replacement of petroleum-based feed-stocks by those from renewable sources. There have been strong efforts in the development and marketing of triglyceride-based polyol technology worldwide from academia to industries. Since this technology can be used in polyurethane applications and requires less energy and will have lower market prize compared to petro-based polyols thereby it has attracted the attention of many multi-national industries and international reputed research groups. Among the triglyceride-based polyols technologies, the soybean oil-based polyol technology is given much more focused. Since soy-based polyols can be applied in many polyurethane formulations mainly in polyurethane foams [16(a)]. Cargill's soy-based polyols are made by conventional chemical methods where carbon-carbon double bonds in unsaturated vegetable oils are converted to epoxy derivatives and then to secondary hydroxyl groups to obtain polyols [6, 7]. Dow chemical company introduced RENUVA™ renewable resource technology to produce biobased polyols. Here, triglycerides are modified by hydroformylation and followed by hydrogenation to give primary hydroxyl groups [8, 9]. Similar to Cargill, Dow Chemical biobased polyols are used for polyurethane formulations to make flexible foams in furniture, bedding, automotive, carpeting, coatings, adhesives, sealants, and elastomers [25]. Bayer Material science (BMS) is also developing natural triglyceride based polyols by adopting chemical methods. Bayer claims that the resulting polyol products have a high content of vegetable oil with low hydroxyl functionality and relatively high molecular weight, which makes the polyols suitable to produce solid urethanes and flexible foams [26]. Vertellus Performance Materials developed a polyol family based on castor-oils to prepare a wider range of polyurethane coatings, sealants, and adhesives. Unlike other vegetable-based polyols that require chemical modification of unsaturated fatty acids to introduce hydroxyl groups, this product is produced directly from a plant resource that contains 85-90% ricinoleic acid, and hydroxyl groups are naturally available into these polyols [26]. There are many research groups such as Dawn Isa *et al* [27], Luo *et al* [28] have similarly reported the chemical methods to prepare bio-based polyols with high hydroxyl numbers (200-500 mg KOH/g) and intermediate viscosities (260-5300 cP at 25 °C).

DRAWBACKS OF EXITING POLYOL TECHNOLOGY

As discussed earlier that polyurethanes are prepared *via* conventional chemical methods using perto-based polyols, and besides that, polyurethane formulations require a high proportion of volatile organic solvents such as DMF, MEK, acetone, toluene, para-xylene, *etc*. The tremendous use of petro-based polyols in polyurethanes and a huge amount of released VOC content pose a threat to the environment, and cause pollution and severe health problems.

The majority of current industrial approaches to produce bio-based polyols for polyurethane manufacturing have one or more drawbacks as discussed above. Among the most active biobased polyols technologies, the development and marketing of soy-based polyol technology are mainly focused. Soy-based polyols are applicable in many polyurethane formulations mainly in polyurethane foams, and it is capable to replace 100% petrochemical-based polyols [29]. However, there are some major drawbacks of soy-based polyols to be considered; (i) Multiple chemical steps, use of many chemicals and solvent increase the production costs. (ii) Availability of only secondary hydroxyl groups that are less reactive compared to primary hydroxyl groups. (iii) Unreacted saturated fatty acid content (10-15% saturated fatty acid in original soybean oil) results in dangling after the reaction with isocyanates and leads to poor polyurethane properties [6].

It is noteworthy to mention that polyols as precursors for polyurethanes are prepared in general by chemical catalyzed polycondensation reactions using organometallic catalysts such as dibutyl tin oxide, acetates of zinc, magnesium, calcium, titanium oxides at high temperatures ($\geq$150 °C) [30, 31, 32a-d, 33a-c]. High-temperature condensation reaction conditions lead to undesirable side reactions, such as dehydration of diol and β-scission of poly-esters to form acid and alkene end groups. These harsh conditions are particularly problematic when using thermally or chemically unstable monomers containing siloxane, epoxy, and vinyl moieties. Besides temperature concerns, residual metals present in catalysts are difficult to remove and can pollute the environment severely upon disposal.

PREPARATION OF BIOCATALYSED POLYOLS

The first lipase catalyzed polymerization using aliphatic diacids and diols was reported by Okumara *et al* in 1984 [34]. In general, step-growth polycondensation and ring-opening polymerizations are the most common processes for lipase catalyzed polyester synthesis. Lipases are the most efficient biocatalysts and extensively studied in enzymatic polymerization to prepare polyesters [35, 36].

Lipase catalysis is also able to convert natural glycerol and sorbitol, xylitol, mannitol, and many more which have multiple hydroxyl groups into polyester polyols as prepolymers. Chemical methods can be used to prepare polyol polyesters but these methods result in uncontrolled structures due to their lack of selectivity. The advantage of biocatalytic routes to provide a linear structure of polyester polyols which have more exposure for reactions with isocyanates. Biocatalysis also enables the synthesis of prepolymers at low reaction temperature as compared to chemical methods which require higher reaction temperature, and therefore more consumption of energy input.

By using lipase-catalyzed condensation reactions where hydroxyl groups are determined by the content of glycerol, sorbitol, or an alternative reduced sugar molecule, the degree of hydroxylation can be varied widely. Furthermore, the rigidity of polyol-polyesters will be a function of building blocks used. It is envisioned that many different grades of polyol-polyesters can be made by this approach where the chain length, degree of hydroxylation, and chain rigidity can be systematically varied. Polyol-polyesters synthesized by enzyme-catalysis in combination with isocyanates provide a new route to polyurethanes that can be used in flexible as well as stiff and rigid polyurethane foams, polyurethane-based coatings, adhesives, sealants, and elastomers. Reduced sugar molecules are readily renewable, inexpensive, and harmless to the environment. Hu *et al.* [37] reported the preparation of a family of polyols from natural alditols. Molecular weight increase as a function of time was used to assess differences in polyol reactivity. The relative weight-average molecular weights of resulting polyol-polyester increases with as a function of reaction time and polyol structure. They have different chain lengths and/or stereochemistry. Given the inherent selectivity of enzymes, it was anticipated that N435-catalyzed polyesterification of these substrates would occur at different rates. In another study by Kulshrestha *et al.* [38], the polymerizations were performed using a 1:1 molar feed ratio of adipic acid to 1,8-octane diol plus glycerol, the condensation polymerizations were catalyzed by N435. CAL-B (Lipase B from *Candida antartica*) catalyzed synthesis of glycerol copolyesters, products formed at selected reaction times during the polymerization were isolated and analyzed. For the systematic study, the monomer feed ratio 100:80:20 (A:O:G) was selected. The polymerization after

5, 20, 30, and 90 min were studied by ^{1}H NMR. During the first 5 min of the polymerization, only OA units were formed, and after 20 and 90 min, the content of GA units increased to 13 and 21 mol %, respectively. The product isolated after 42 h has 18 mol % GA units. Thus, adipic acid reacted more rapidly with 1,8-octane diol than glycerol. An inverse gated ^{13}C NMR spectrum of the product formed after 2 h showed signals corresponding to un-reacted glycerol were absent. By using lipase-catalyzed condensation reactions where prepolymer hydroxyl group content is determined by the amount of glycerol and sorbitol in the monomer feed. Furthermore, polyol rigidity is a main function of the building blocks used instead of being restricted to the basic characteristics of triglycerides. Lipase-catalyzed condensation reactions can be performed in bulk (without solvent) at moderate temperatures that reduce energy consumption. Furthermore, prepolymer synthesis does not generate toxic by-products and it is performed in one-pot. Moreover, lipase-catalysis can lead to selective condensation reactions giving nearly linear prepolymers and narrow prepolymer polydispersity (*e.g.* M_w/M_n=1.5).

Multiple-functional monomers such as glycerol and sugar alcohols can be polymerized directly *via* enzymatic polymerization due to the high regio-selectivity of enzymes. Due to the high regio-selectivity of N-435, the enzymatic polymerizations of glycerol, adipic acid and octane diol gave linear polyesters at short reaction time but it leads to highly branched structures at a long reaction time. Here, N-435 showed 77 to 82% of the regioselectivity towards primary alcohols in such a polyesterification reaction [38 - 40]. Similarly, the lipase catalyzed polymerization with D-sorbitol was also studied, and it also indicates that the primary hydroxyl group of D-sorbitol exclusively esterified during the enzymatic polymerization. NMR studies reveal that N-435 showed high regioselectivity towards the primary hydroxyl group of D-sorbitol [41].

Lipase catalysis has also been used in ring-opening polymerization of lactones, as it was reported by the research group of Gutman and Kobayashi [42, 43]. Gutman *et al.* successfully carried out the ring-opening polymerization of caprolactone in hexane using lipase as a catalyst, and the resulting polymer was successfully obtained with Mn = 4400. Kobayashi et. al performed the successful enzymatic ring-opening polymerization of lactones in bulk using different lipases as catalysts.

Further, many research groups have successfully reported the ring-opening polymerization of various types of lactones using CALB as catalyst [44a-f]. Cyclic ε-caprolactone (ε-CL) can be readily polymerized by most of the lipases. Among lipase families, CALB has been widely used both in bulk and ionic liquids for eROP of ε-CL [45a-d].

The same has been studied by Deng *et al.*, and reported 80% yield of the resulting products in bulk with a small amount of immobilized enzyme in 4 hours [46]. Another lipase Pseudomonas fluorescens(PF) in non-immobilized form has also been used in ring-opening polymerization of δ-Valerolactone [47]. Frey *et al.* [48] reported a method in which ROP of the lactone combines with polycondensation of the carboxylic acid using N-435 as a catalyst, to prepare a series of hyperbranched copolyesters with different degrees of branching. The preparation of such hyperbranched polymers can be carried out in solution and bulk conditions.

Various combinations of monomers such as diacids/diesters diols and polyols, hydroxyacids/esters, and cyclic monomers like lactones, cyclic diesters, and cyclic ketene acetals, have been studied for the lipase-catalyzed polymerization. The progress of these pioneer works has been summarized in various review articles [35, 49a-f, 50].

The majority of enzymes studied for polymerization reaction such as step condensation reaction, transesterification, and ring-opening polymerization have been from the lipase family and particularly with lipase B from *Candida antartica*. Cutinases from *Hemicola insolens* have also been reported for catalyzing polyester synthesis [51]. Cutinases are extracellular fungal enzymes whose ability is to catalyze the hydrolysis of cutin [52]. Many research papers on cutinase-catalyzed biotransformation have mainly been focused on polyester hydrolysis [53] and esterification reaction between small molecules [54]. As similar to lipase, cutinase *Hemicola insolens* has shown remarkable activity for ring-opening and condensation polyester reaction [51].

The above-discussed enzyme-catalyzed polymerization of varied monomer substrates may lead to polyester polyols prepolymers which act as precursors for polyurethanes.

As commercial development progresses, polyurethanes can be synthesized from bio-based prepolymers with a range of commercially desired mechanical properties. This can be accomplished by variation in prepolymer chain length, rigidity, and degree of hydroxylation.

Scheme to Prepare Polyol polyesters

Synthesis of polyol polyesters containing biobased building blocks such as glycerol/sorbitol/sebacic acid/furan dicarboxylic acid can be achieved *via* immobilized lipase-catalysis as given below in the Scheme (**4**).

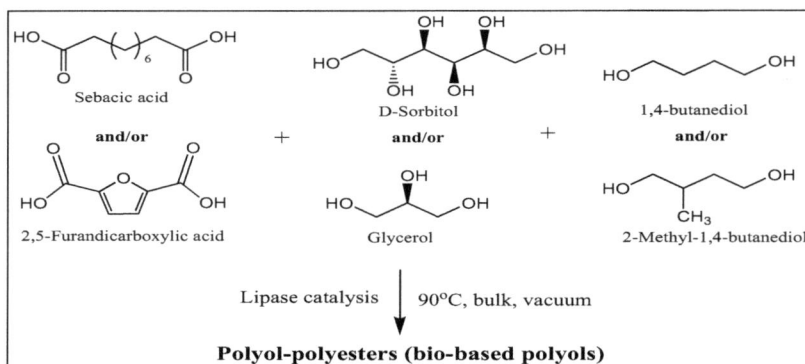

Scheme (4). Polyol polyesters synthesis by lipase-catalysis.

Structure of Expected Polyol Polyesters

1. Polycondensation reaction between Sebacic acid, glycerol, and 1,4-butanediol(with a varied molar ratio of glycerol and butanediol against sebacic acid) may result in the following structure of polyol polyesters:

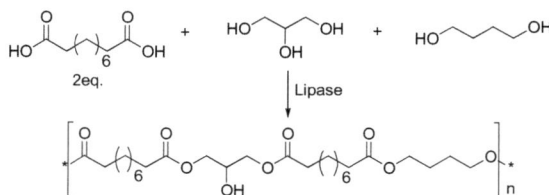

2. Polycondensation reaction between Sebacic acid, D-sorbitol, and 2-methyl-1-4-butanediol (with a varied molar ratio of D-sorbitol and glycerol against sebacic acid) may result in the following structure of Polyol polyesters

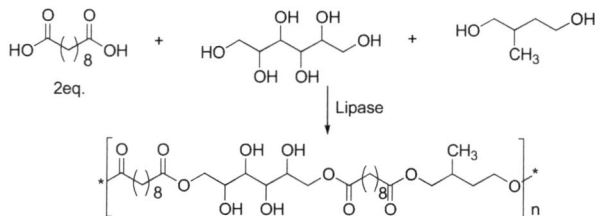

3. The reaction between 2,5-Furandicarboxylic acid, Glycerol, and 1,4-butanediol (with a varied molar ratio of Glycerol and 1,4-butanediol against sebacic acid) may result in the following structure of Polyol polyesters:

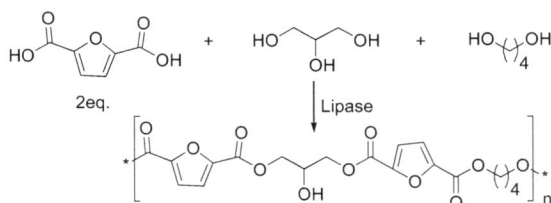

Biobased Building Blocks For Polyol Polyester Synthesis

Biomass feedstocks have been used extensively to produce various types of monomers *via* bio-catalytic and chemo-catalytic processes, which eventually leads to access diverse biobased polymers [55a-f, 56a-c]. Fast development in biotechnology techniques will enable more and more biobased monomers to be commercially available in the market, and their prices will be competitive with that of petrol-derived chemicals [57a-e, 58a-c].

The information below describes availability in manufacturing biobased building blocks to be used to prepare a versatile family of polyol polyester prepolymers for polyurethane synthesis.

Sebacic acid(SA) is a member of oleochemical derived dicarboxylic acids (azelaic, sebacic, and dimer acids) manufactured in 100,000 tons per year as components for polymers [59]. Hengshui Dongfeng Chemical Co.Ltd. claims to be the largest producer of sebacic acid in china (23 000 tons annually). *Furandicarboxylic acid (FDCA)* is formed by oxidative dehydration of glucose [60]. Large scale protocols have been developed to manufacture FDCA that promise to soon make available commodity quantities of these monomer building blocks [61]. *Glycerol* is produced commercially *via* chemical transesterification of triglycerides [60]. Glycerol production was estimated at USD 1.64billion in 2014 and is expected to grow at about 7.9% till 2020 [62]. *Sorbitol* is produced commercially by hydrogenation of glucose using Raney nickel as a catalyst [60]. ADM is the leading manufacturer of sorbitol with world production levels of about 360 million pounds per year [63]. 1,4-butanediol (BDO) is produced from succinate in a two-step process [64]. Currently, succinate can be produced by fermentation at about \$0.55–1.10 per kg. As an alternate of petroleum-based chemicals, commercial facilities for the fermentative production of succinate and succinate-derived BDO are emerging. For example, Mitsubishi Chemical (Japan) had a plan to make enough bio-based succinate to produce 30,000 Mt/year of biodegradable polybutylene succinate [65]. Also, the Research Institute of Innovative Technology for the Earth and Showa Highpolymer has developed a genetically modified strain of *Corynebacterium glutamicum* that produces 50,000 t succinate/year from a wastepaper feedstock [66]. Furthermore, DSM (The Netherlands) and Diversified Natural Products (East Lansing, MI) have a joint venture to commercialize bio-based succinate-derived products [65]. Itaconic acid is a C5 dicarboxylic acid, also known as methylene succinic acid, that is produced *via* fermentation [67, 68] and is a key building block for deriving 2-methyl 1,4-butanediol (2-MeBDO) [60] as well as other specialty chemicals.

ADVANTAGE OF BIOCATLYSIS

There has been a great impetus in the replacement of petroleum-based polyols with biobased polyols in polyurethane applications. However, triglyceride-based polyols have many drawbacks as discussed earlier, and also don't meet adequate physical performance characteristics that restrict them to limited uses.

In contrast, biocatalysis has inherent advantages for the preparation of a wide range of functional polymers due to mild reaction conditions, high tolerance of enzymes for functional groups, and catalyst selectivity that provides better control over branching.

There are many esteemed research groups and research centers worldwide those are pioneers and highly engaged in enzyme-catalysis research. Research group of Gross and Kobyashi have reported the successful enzyme-catalyzed syntheses of polyester polyols, which were achieved by condensation copolymerization of an aliphatic diacid with glycerol or sorbitol, terpolymerization of aliphatic diacids and diols with different types of carbohydrate molecules [37, 40]. Most of the polymerization reactions have been performed using immobilized Candida antartica lipase B(CALB) because of its high regio-, chemo- and enantio-selectivity and its high thermal stability and activity. Similar studies have been performed by other research groups including Varma *et al.* [69], Kelly *et al.* [70], Klibanov *et al.* [71], Dordick *et al.* [72] Hult *et al.* [73], Kirk *et al.* [74], and Ballesteros *et al.* [75], Balcão *et al.* [76], etc, and have published outstanding research articles and review articles demonstrating the diverse application of enzyme catalysis in general and lipase Catalysis in organic synthesis including many regio and enantio-selective syntheses. Furthermore, CALB shows a wide range of applications that have been developed not only within academic research but also for pilot scale manufacturing of pharmaceuticals, chemicals, monomer, and polymers. Furthermore, by using a solvent-free biocatalytic process, biobased building blocks, and moderate reaction temperatures, products can be developed that are safe, clean, and energy-efficient. Also, the enzymatic routes provide better control over branching, and thereby avoiding cross-linking unlike chemical catalysis [50, 77a-c]. Also, lipase-catalysis can lead to selective condensation reactions giving nearly linear prepolymers. This has the important advantage to avail all hydroxyls along the prepolymer highly accessible for reactions with isocyanates. Polyols with variable properties can be accomplished by variation in prepolymer chain length, rigidity, and degree of hydroxylation. Furthermore, the polyols can be separated from catalyst beads by decantation or using a filtration step. This will permit catalysts to use multiple times thereby lowering manufacturing costs. Thus, bio-catalyzed polyols are expected to be used in manufacturing stiff and flexible polyurethane foams, polyurethane-based coatings, adhesives, sealants, elastomers, *etc.*

SIGNIFICANCE OF BIOCATALYSIS

Biocatalysis, in combination with biorenewable monomers, can be employed to develop liquid polyols for the preparation of solvent-less borne polyurethanes. From an economic and environmental point of view, it is a great deal and opportunity to replace a substantial fraction of petroleum-based materials with bio-based materials from renewable resources synthesized by the environmentally friendly and green biocatalytic route.

The scheme described above may help in developing such methods and processes that: (i) require less energy and low cost, (ii) minimize the use of toxic chemicals, (iii) decrease the generation of toxic by-products, iv) minimal emission of VOCs, v) incorporate safe renewable feedstocks into products.

Biocatalysis is the key to the sustainable development of chemicals and materials. The use of biorenewable materials and enzyme-catalysis lead to may immeasurable benefits to the nation and society as follows.

i. The use of biorenewbale materials would reduce our dependence on petro-based materials.
ii. Tremendous consumption of biorenewbale materials will help agricultural sectors to boom, and to create new markets for grain and crop residue, and help stimulate agricultural economies.
iii. Enzyme-catalysis requires low temperature and mild reaction conditions thereby low energy consumption.

The biocatalytic routes to develop biobased polyols as a potential precursor for polyurethane manufacturing would help in reducing the use of toxic chemicals, decrease the generation of toxic by-products, minimal emission of VOCs, *etc*, and thereby keep the environment, society, and nation safe and healthy.

CONSENT FOR PUBLICATION

Not applicable.

CONFLICT OF INTEREST

The authors confirm that there is no conflict of interest.

ACKNOWLEDGEMENTS

B Sharma is grateful to Prof. R. A. Groos, (moved from the Polytechnic University of NYU, NY to Rensselaer Polytechnic Institute, NY) for his supervision, continuous support and encouragement, and the knowledge he has gained about the green chemistry and biocatalysis during postdoctoral work at Polytechnic University of NYU, NY.

REFERENCES

[1]   Oertel, G. *Polyurethane Handbook,* 2nd ed; Hanser Publishers, **1993**.

[2]   https://www.grandviewresearch.com/industry-analysis/polyurethane-pu-market

[3]   https://www.grandviewresearch.com/industry-analysis/polyols-market

[4]   Mowre, N.R. Solventless polyurethane spray compositions and method for applying them. US Patent 4695618.

[5]   Hong, C.H. Solvent-free polyurethane-based artificial leather having the texture of human skin and the preparation method thereof. US Patent Application 2009/0247671A1 2009.

[6]   Zlantanic, A.; Lava, C.; Zhang, W.; Petrovic, Z.S. Effect of structure on properties of polyols and polyurethanes based on different vegetable oils. *J. Polym. Sci., B, Polym. Phys.,* **2004**, *42*, 809-819. [http://dx.doi.org/10.1002/polb.10737]

[7]   Guo, A.; Cho, Y-J.; Petrovic, Z.S. Structure and properties of halogenated and nonhalogenated soy☐based polyols. *J. Polym. Sci. A Polym. Chem.,* **2000**, *38*, 3900-3910. [http://dx.doi.org/10.1002/1099-0518(20001101)38:21<3900::AID-POLA70>3.0.CO;2-E]

[8]   Guo, A.; Demydov, D.; Zhang, W.; Petrovic′, Z.S. Polyols and Polyurethanes from Hydroformylation of Soybean Oil. *J. Polym. Environ.,* **2002**, *10*, 49-52. [http://dx.doi.org/10.1023/A:1021022123733]

[9]   Kandanarachchi, P.; Guo, A.; Petrovic′, Z.S. The hydroformylation of vegetable oils and model compounds by ligand modified rhodium catalysis. *J. Mol. Catal. Chem.,* **2002**, *184*, 65-71. [http://dx.doi.org/10.1016/S1381-1169(01)00420-4]

[10]   Petrović, Z.S.; Zhang, W.; Javni, I. Structure and properties of polyurethanes prepared from triglyceride polyols by ozonolysis. *Biomacromolecules,* **2005**, *6*(2), 713-719. [http://dx.doi.org/10.1021/bm049451s] [PMID: 15762634]

[11]   Tran, P.; Graiver, D.; Narayan, R. Ozone-mediated polyol synthesis from soybean oil. *J. Am. Oil Chem. Soc.,* **2005**, *82*, 653-659. [http://dx.doi.org/10.1007/s11746-005-1124-z]

[12]   Petrovic´, Z.S.; Zlantanic, A.; Lava, C.C.; Sinadinovic-Fiser, S. Epoxydation of Soybean Oil in Toluene with Peroxoacetic acid and Peroxoformic acid- Kinetics and side reactions. *Eur. J. Lipid Sci. Technol.,* **2002**, *104*, 293-299.
[http://dx.doi.org/10.1002/1438-9312(200205)104:5<293::AID-EJLT293>3.0.CO;2-W]

[13]   Petrovic´, Z.S. Polyurethanes from Vegetable Oils. *Polym. Rev. (Phila. Pa.),* **2008**, *48*, 109-155.
[http://dx.doi.org/10.1080/15583720701834224]

[14]   Abraham, T. http://www.cargill.com/company/businesses/bioh-polyols/index.jsp

[15]   Abraham, T. Cargill at IPRIME meeting, Univ of Minnesota May. **2006**.

[16]   a) Frankel, E.N. Catalytic Hydroformylation of Unsaturated Fatty Derivatives with Cobalt Carbonyl. *J. Am. Oil Chem. Soc.,* **1976**, *53*, 138-141.
[http://dx.doi.org/10.1007/BF02586351] b) Frankel, E.N.; Thomas, F.L. Selective Hydroformylation of Polyunsaturated Fats With a Rhodium-Triphenylphosphine Catalyst. *J. Am. Oil Chem. Soc.,* **1972**, *49*, 10-14.
[http://dx.doi.org/10.1007/BF02545129] c) Guo, A.; Demydov, D.; Zhang, W.; Petrovic´, Z.S. Polyols and Polyurethanes from Hydroformylation of Soybean Oil. *J. Polym. Environ.,* **2002**, *10*, 49-52.
[http://dx.doi.org/10.1023/A:1021022123733] d) Kandanarachchi, P.; Guo, A. Kinetics of the Hydroformylation of Soybean Oil by Ligand-Modified Homogeneous Rhodium Catalysis. *J. Am. Oil Chem. Soc.,* **2002**, *79*, 1221-1225.
[http://dx.doi.org/10.1007/s11746-002-0631-2] e) Kandanarachchi, P.; Guo, A.; Petrovic´, Z.S. The hydroformylation of vegetable oils and model compounds by ligand modified rhodium catalysis. *J. Mol. Catal. Chem.,* **2002**, *184*, 65-71.
[http://dx.doi.org/10.1016/S1381-1169(01)00420-4]

[17]   http://www.dow.com/polyurethane/news/2007/20070925a.htm

[18]   Zlantanic, A.; Petrovic´, Z.S.; Dušek, K. Structure and Properties of Triolein-Based Polyurethane Networks. *Biomacromolecules,* **2002**, *3*, 1048-1056.
[http://dx.doi.org/10.1021/bm020046f] [PMID: 12217052]

[19]   Ionescu, M. *Chemistry and Technology of Polyols for Polyurethanes*; Rapra Technology Limited UK, **2005**.

[20]   Petrović, Z.S.; Zhang, W.; Javni, I. Structure and properties of polyurethanes prepared from triglyceride polyols by ozonolysis. *Biomacromolecules,* **2005**, *6*(2), 713-719.
[http://dx.doi.org/10.1021/bm049451s] [PMID: 15762634]

[21]   Tran, P.; Graiver, D.; Narayan, R. Ozone-Mediated Polyol Synthesis from Soybean Oil. *J. Am. Oil Chem. Soc.,* **2005**, *82*, 653-659.
[http://dx.doi.org/10.1007/s11746-005-1124-z]

[22]   a) Dawn'Isa, J.-P. L.; Drzal, L. T.; Mohanty, A. K.; Misha, M.,. US Patent 7125950B2. b) Nieman, L.K. European Patent Application 1792926A2.

[23]   Ionescu, M.; Petrovic´, Z.S.; Wan, X. Ethoxylated Soybean Polyols for Polyurethanes. *J. Polym. Environ.,* **2007**, *15*, 237-243.
[http://dx.doi.org/10.1007/s10924-007-0065-4]

[24]   Luo, N.; Newbold, T. Method for vegetable oil derived polyols and polyurethanes made therefrom. US Patent Application 0262259.

[25]   http://www.dow.com/polyurethane/news/2007/20070925a.htm

[26]   http://www.ptonline.com/articles/200712fa4.html

[27]   Dawn'Isa, J-P. L.; Drzal, L. T.; Mohanty, A. K.; Misha, M. US patent 7125950B2.

[28]   Luo, N.; Newbold, T. Method for vegetable oil derived polyols and polyurethanes made therefrom. US Patent Application 2008/0262259. 2008.

[29]   www.cargill.com

[30]   Stumebe, J.F.; Bruchmann, B. Hyperbranched Polyesters Based on Adipic Acid and Glycerol. *Macromol. Rapid Commun.,* **2004**, *25,* 921-924.
[http://dx.doi.org/10.1002/marc.200300298]

[31]   Gross, R.A.; Ganesh, M.; Lu, W. Enzyme-catalysis breathes new life into polyester condensation polymerizations. *Trends Biotechnol.,* **2010**, *28*(8), 435-443.
[http://dx.doi.org/10.1016/j.tibtech.2010.05.004] [PMID: 20598389]

[32]   a) Gao, C.; Yan, D. Hyperbranched polymers: from synthesis to applications. *Prog. Polym. Sci.,* **2004**, *29*, 183-275.
[http://dx.doi.org/10.1016/j.progpolymsci.2003.12.002] b) Flory, P.J. *Principles of Polymer Chemistry*; Cornell University Press: Ithaca, NY, **1953**. c) Kim, Y.H. Hyperbranched Polymers 10 Years After. *J. Polym. Sci. A Polym. Chem.,* **1998**, *36*, 1685-1698.
[http://dx.doi.org/10.1002/(SICI)1099-0518(199808)36:11<1685::AID-POLA1>3.0.CO;2-R] d) Hult, A.; Johansson, M.; Malmstro¨, E. Hyperbranched Polymers. *Adv. Polym. Sci.,* **1999**, *143*, 1-34.
[http://dx.doi.org/10.1007/3-540-49780-3_1]

[33]   a) Brioudea, M.M.; Guimarãesa, D.H.; Boaventura, J.S.; José, N.M. Synthesis and Characterization of Aliphatic Polyesters from Glycerol, by-Product of Biodiesel Production, and Adipic Acid. *Mater. Res.,* **2007**, *10*, 335-339.
[http://dx.doi.org/10.1590/S1516-14392007000400003] b) Wyatt, V.T.; Nunez, A.; Foglia, T.A.; Marmer, W.N. Synthesis of HyperbranchedPoly(glycerol-diacid) Oligomers. *J. Am. Oil Chem. Soc.,* **2006**, *86*, 1033-1039.
[http://dx.doi.org/10.1007/s11746-006-5159-y] c) Wang, Y.; Ameer, G.A.; Sheppard, B.J.; Langer, R. A tough biodegradable elastomer. *Nat. Biotechnol.,* **2002**, *20*(6), 602-606.
[http://dx.doi.org/10.1038/nbt0602-602] [PMID: 12042865]

[34]   Okumura, S.; Iwai, M.; Tominaga, Y. Synthesis of ester oligomer by Aspergillus niger lipase. *Agric. Biol. Chem.,* **1984**, *48*, 2805-2808.

[35]   Mileti'c, N.; Loos, K.; Gross, R.A. Enzymatic polymerization of polyester.*Biocatalysis in Polymer Chemistry*; Loos, K., Ed.; Wiley-VCH Verlag GmbH & Co. KGaA: Weinheim, Germany, **2010**, pp. 83-129.

[36]   Kobayashi, S. Enzymatic ring-opening polymerization and polycondensation for the green synthesis of polyesters. *Polym. Adv. Technol.,* **2015**, *26*, 677-686.
[http://dx.doi.org/10.1002/pat.3564]

[37]   Hu, J.; Gao, W.; Kulshrestha, A.; Gross, R.A. "Sweet Polyesters": Lipase-Catalyzed Condensation-Polymerizations of Alditols. *Macromolecules,* **2006**, *39*, 6789-6792.
[http://dx.doi.org/10.1021/ma0612834]

[38]   Kulshrestha, A.S.; Gao, W.; Gross, R.A. Glycerol Copolyesters: Control of Branching and Molecular Weight Using a Lipase Catalyst. *Macromolecules,* **2005**, *38*(8), 3193-3204.
[http://dx.doi.org/10.1021/ma0480190]

[39]   Fu, H.; Kulshrestha, A.S.; Gao, W.; Gross, R.A.; Baiardo, M.; Scandola, M. Physical Characterization of Sorbitol or Glycerol Containing Aliphatic Copolyesters Synthesized by Lipase-Catalyzed Polymerization. *Macromolecules,* **2003**, *36*, 9804-9808.
[http://dx.doi.org/10.1021/ma035129i]

[40]   Kumar, A.; Kulshrestha, A.; Gao, W.; Gross, R.A. Versatile Route to Polyol Polyesters by Lipase Catalysis. *Macromolecules,* **2003**, *36*, 8219-8221.
[http://dx.doi.org/10.1021/ma0351827]

[41]   Uyama, H.; Klegraf, E.; Wada, S.; Kobayashi, S. Regioselective Polymerization of Sorbitol and Divinyl Sebacate Using Lipase Catalyst. *Chem. Lett.,* **2000**, *29*, 800-801.
[http://dx.doi.org/10.1246/cl.2000.800]

[42]   Knani, D.; Gutman, A.L.; Kohn, D.H. Enzymatic Polyesterification in Organic Media. Enzyme
       Catalyzed Synthesis of Linear Polyesters. 1. Condensation Polymerization of linear Hydroxyesters. II.
       Ring-Opening Polymerization of ε-Caprolactone. *J. Polym. Sci. A Polym. Chem.,* **1993**, *31*, 1221-
       1232.
       [http://dx.doi.org/10.1002/pola.1993.080310518]

[43]   Uyama, H.; Kobayashi, S. Enzymatic Ring-Opening Polymerization of Lactones Catalyzed by Lipase.
       Chem. Lett., **1993**, *22*, 1149-1150.
       [http://dx.doi.org/10.1246/cl.1993.1149]

[44]   a) Matsumura, S. Enzymatic Synthesis of Polyesters via Ring-Opening Polymerization. *Adv. Polym.
       Sci.,* **2006**, *194*, 95-132.
       [http://dx.doi.org/10.1007/12_030] b) Binns, F.; Harffey, P.; Roberts, S.M.; Taylor, A. Studies of
       Lipase-Catalyzed Polyesterification of an Unactivated Diacid/Diol System. *J. Polym. Sci. A Polym.
       Chem.,* **1998**, *36*, 2069-2079.
       [http://dx.doi.org/10.1002/(SICI)1099-0518(19980915)36:12<2069::AID-POLA13>3.0.CO;2-4]   c)
       Dong, H.; Cao, S.G.; Li, Z.Q.; Han, S.P.; You, D.L.; Shen, J.C. Study on the Enzymatic
       Polymerization Mechanism of Lactone and the Strategy for Improving the Degree of Polymerization.
       J. Polym. Sci. A Polym. Chem., **1999**, *37*, 1265-1275.
       [http://dx.doi.org/10.1002/(SICI)1099-0518(19990501)37:9<1265::AID-POLA6>3.0.CO;2-I] Uyama,
       H.; Kobayashi, S. *In Enzyme-Catalyzed Synthesis of Polymers*; Springer: New York, **2006**, Vol. 194,
       pp. 133-158.
       [http://dx.doi.org/10.1007/12_031] e) Kumar, A.; Gross, R.A. Candida antartica lipase B catalyzed
       polycaprolactone synthesis: effects of organic media and temperature. *Biomacromolecules,* **2000**, *1*(1),
       133-138.
       [http://dx.doi.org/10.1021/bm990510p] [PMID: 11709835] f) Thurecht, K.J.; Heise, A.; deGeus, M.;
       Villarroya, S.; Zhou, J.X.; Wyatt, M.F.; Howdle, S.M. Kinetics of Enzymatic Ring-Opening
       Polymerization of ε-Caprolactone in Supercritical Carbon Dioxide. *Macromolecules,* **2006**, *39*, 7967-
       7972.
       [http://dx.doi.org/10.1021/ma061310q]

[45]   a) Mei, Y.; Kumar, A.; Gross, R. Kinetics and Mechanism of Candida antarctica Lipase B Catalyzed
       Solution Polymerization of e-Caprolactone. *Macromolecules,* **2003**, *36*, 5530.
       [http://dx.doi.org/10.1021/ma025741u] Heise, A.; van der Meulen, I. *Green Polymerization Methods*;
       Wiley-VCH: Weinheim, Germany, **2011**, p. 291.
       [http://dx.doi.org/10.1002/9783527636167.ch13] c) Uyama, H.; Takamoto, T.; Kobayashi, S.
       Enzymatic Synthesis of Polyesters in Ionic Liquids. *Polym. J.,* **2002**, *34*, 94.
       [http://dx.doi.org/10.1295/polymj.34.94] d) Yuan, J.; Dai, Y.; Yu, Y.; Wang, P.; Wang, Q.; Fan, X.
       Biocatalytic synthesis of poly(ε-caprolactone) using modified lipase in ionic liquid media. *Eng. Life
       Sci.,* **2016**, *16*, 371.
       [http://dx.doi.org/10.1002/elsc.201500081]

[46]   Deng, F.; Gross, R.A. Ring-opening bulk polymerization of epsilon-caprolactone and trimethylene
       carbonate catalyzed by lipase Novozym 435. *Int. J. Biol. Macromol.,* **1999**, *25*(1-3), 153-159.
       [http://dx.doi.org/10.1016/S0141-8130(99)00029-X] [PMID: 10416662]

[47]   Kobayashi, S.; Makino, A.; Matsumoto, H.; Kunii, S.; Ohmae, M.; Kiyosada, T.; Makiguchi, K.;
       Matsumoto, A.; Horie, M.; Shoda, S. Enzymatic polymerization to novel polysaccharides having a
       glucose-N-acetylglucosamine   repeating   unit,   a   cellulose-chitin   hybrid   polysaccharide.
       Biomacromolecules, **2006**, *7*(5), 1644-1656.
       [http://dx.doi.org/10.1021/bm060094q] [PMID: 16677050]

[48]   Skaria, S.; Smet, M.; Frey, H. Enzyme-Catalyzed Synthesis of Hyperbranched Aliphatic Polyesters.
       Macromol. Rapid Commun., **2002**, *23*, 292.
       [http://dx.doi.org/10.1002/1521-3927(20020301)23:4<292::AID-MARC292>3.0.CO;2-5]

[49]   a) Kobayashi, S.; Uyama, H.; Kimura, S. Enzymatic polymerization. *Chem. Rev.,* **2001**, *101*(12),
       3793-3818.

[http://dx.doi.org/10.1021/cr990121l] [PMID: 11740921] b) Kobayashi, S.; Makino, A. Enzymatic polymer synthesis: an opportunity for green polymer chemistry. *Chem. Rev.,* **2009**, *109*(11), 5288-5353.
[http://dx.doi.org/10.1021/cr900165z] [PMID: 19824647] c) Gross, R.A.; Ganesh, M.; Lu, W. Enzyme-catalysis breathes new life into polyester condensation polymerizations. *Trends Biotechnol.,* **2010**, *28*(8), 435-443.
[http://dx.doi.org/10.1016/j.tibtech.2010.05.004] [PMID: 20598389] d) Kobayashi, S. Enzymatic ring-opening polymerization and polycondensation for the green synthesis of polyesters. *Polym. Adv. Technol.,* **2015**, *26*, 677-686.
[http://dx.doi.org/10.1002/pat.3564] e) Yu, Y.; Wu, D.; Liu, C.B.; Zhao, Z.H.; Yang, Y.; Li, Q.S. Lipase/esterase-catalyzed synthesis of aliphatic polyesters via polycondensation: A review. *Process Biochem.,* **2012**, *47*, 1027-1036.
[http://dx.doi.org/10.1016/j.procbio.2012.04.006] f) Shoda, S.; Uyama, H.; Kadokawa, J.; Kimura, S.; Kobayashi, S. Enzymes as Green Catalysts for Precision Macromolecular Synthesis. *Chem. Rev.,* **2016**, *116*(4), 2307-2413.
[http://dx.doi.org/10.1021/acs.chemrev.5b00472] [PMID: 26791937]

[50]   Gross, R.A.; Kumar, A.; Kalra, B. Polymer synthesis by in vitro enzyme catalysis. *Chem. Rev.,* **2001**, *101*(7), 2097-2124.
[http://dx.doi.org/10.1021/cr0002590] [PMID: 11710242]

[51]   Mo, H.; Azim, A.; Mang, H.; Wallner, S.R.; Ronkvist, A.; Xie, W.; Gross, R.A. A Cutinase with Polyester Synthesis Activity. *Macromolecules,* **2007**, *40*, 148-150.
[http://dx.doi.org/10.1021/ma062095g]

[52]   Heredia, A. BBA-Gen. *Subjects,* **2003**, *1620*, 1-7.
[http://dx.doi.org/10.1016/S0304-4165(02)00510-X]

[53]   Guebitz, G.M.; Cavaco-Paulo, A. Enzymes go big: surface hydrolysis and functionalization of synthetic polymers. *Trends Biotechnol.,* **2008**, *26*(1), 32-38.
[http://dx.doi.org/10.1016/j.tibtech.2007.10.003] [PMID: 18037176]

[54]   Carvalho, C.M.L. Cutinase structure, function and biocatalytic applications. *Electron. J. Biotechnol.,* **1998**, *1*, 160-173.
[http://dx.doi.org/10.2225/vol1-issue3-fulltext-8]

[55]   a) Isikgor, F.H.; Remzi Becer, C. Lignocellulosic Biomass: A Sustainable Platform for Production of Bio-Based Chemicals and Polymers. *Polym. Chem.,* **2015**, *6*, 4497-4559.
[http://dx.doi.org/10.1039/C5PY00263J] b) Gandini, A. Monomers and macromonomers from renewable resources. In Biocatalysis in Polymer ChemistryWiley-VCH Verlag GmbH & Co. KGaA: Weinheim, Germany, **2010**, pp. 1-33.c) Gandini, A.; Lacerda, T.M. From monomers to polymers from renewable resources: Recent advances. *Prog. Polym. Sci.,* **2015**, *48*, 1-39.
[http://dx.doi.org/10.1016/j.progpolymsci.2014.11.002] d) Gandini, A.; Lacerda, T.M.; Carvalho, A.J.; Trovatti, E. Progress of Polymers from Renewable Resources: Furans, Vegetable Oils, and Polysaccharides. *Chem. Rev.,* **2016**, *116*(3), 1637-1669.
[http://dx.doi.org/10.1021/acs.chemrev.5b00264] [PMID: 26291381] e) Dove, A. From monomers to polymers from renewable resources: Recent advances. *Science,* **2012**, *335*, 1382-1384.
[http://dx.doi.org/10.1126/science.335.6074.1382] f) Gallezot, P. Conversion of biomass to selected chemical products. *Chem. Soc. Rev.,* **2012**, *41*(4), 1538-1558.
[http://dx.doi.org/10.1039/C1CS15147A] [PMID: 21909591]

[56]   a) Vilela, C.; Sousa, A.F.; Fonseca, A.C.; Serra, A.C.; Coelho, J.F.J.; Freire, C.S.R.; Silvestre, A.J.D. The quest for sustainable polyesters – insights into the future. *Polym. Chem.,* **2014**, *5*, 3119-3141.
[http://dx.doi.org/10.1039/C3PY01213A] b) Galbis, J.A.; García-Martín, Mde.G.; de Paz, M.V.; Galbis, E. Synthetic Polymers from Sugar-Based Monomers. *Chem. Rev.,* **2016**, *116*(3), 1600-1636.
[http://dx.doi.org/10.1021/acs.chemrev.5b00242] [PMID: 26291239] c) Sousa, A.F.; Vilela, C.; Fonseca, A.C.; Matos, M.; Freire, C.S.R.; Gruter, G-J.M.; Coelho, J.F.J.; Silvestre, A.J.D. Biobased polyesters and other polymers from 2,5- furandicarboxylic acid: a tribute to furan excellency. *Polym. Chem.,* **2015**, *6*, 5961-5983.

[http://dx.doi.org/10.1039/C5PY00686D]

[57]   a) Sheldon, R.A. Green and Sustainable Manufacture of Chemicals from Biomass: State of the Art. *Green Chem.,* **2014**, *16*, 950-963.
[http://dx.doi.org/10.1039/C3GC41935E] b) Becker, J.; Wittmann, C. Advanced Biotechnology: Metabolically Engineered Cells for the Bio-Based Production of Chemicals and Fuels, Materials, and Health-Care Products. *Angew. Chem. Int. Ed.,* **2015**, *54*, 3328-3350.
[http://dx.doi.org/10.1002/anie.201409033] c) Choi, S.; Song, C.W.; Shin, J.H.; Lee, S.Y. Biorefineries for the production of top building block chemicals and their derivatives. *Metab. Eng.,* **2015**, *28*, 223-239.
[http://dx.doi.org/10.1016/j.ymben.2014.12.007] [PMID: 25576747] Dusselier, M.; Mascal, M.; Sels, B.F. Top chemical opportunities from carbohydrate biomass: A chemist's view of the biorefinery.*Selective Catalysis for Renewable Feedstocks and Chemicals*; Nicholas, K.M., Ed.; Springer-Verlag: Berlin/Heidelberg, Germany, **2014**, Vol. 353, pp. 1-40.
[http://dx.doi.org/10.1007/128_2014_544] e) Harmsen, P.F.H.; Hackmann, M.M.; Bos, H.L. Green building blocks for bio-based plastics. *Biofuels Bioprod. Biorefin.,* **2014**, *8*, 306-324.
[http://dx.doi.org/10.1002/bbb.1468]

[58]   Taylor, R.; Nattrass, L.; Alberts, G.; Robson, P.; Chudziak, C.; Bauen, A.; Libelli, I.M.; Lotti, G.; Prussi, M.; Nistri, R. *From the Sugar Platform to Biofuels and Biochemicals; contract No. ENER/C2/423-2012/SI2.673791; E4tech*; RE-CORD and WUR: London, UK, **2015**, pp. 1-183.b) De Jong, E.; Higson, A.; Walsh, P.; Wellisch, M. *Bio-Based Chemicals: Value Added Products from Biorefineries; Avantium Chemicals*; Wageningen, The Netherlands: (Netherlands), NNFCC (UK), Energy Research Group (Ireland), and Agriculture and Agri-Food Canada (Canada), **2012**. Golden, J.; Handfield, R. *Why Biobased? Opportunities in the Emerging Bioeconomy*; US Department of Agriculture, Office of Procurement and Property Management: Washington, DC, USA, **2014**.

[59]   Hill, K. Industrial development and application of biobased oleochemicals. *Pure Appl. Chem.,* **2007**, *79*, 1999-2011.
[http://dx.doi.org/10.1351/pac200779111999]

[60]   Werpy, T.; Petersen, G. Hyperbranched Polyesters Based on Adipic Acid and Glycerol. Macromolecular Rapid Commun. 2003, *25*, 921-924.

[61]   Lichtenthaler, F.W. Unsaturated O- and N-heterocycles from carbohydrate feedstocks. *Acc. Chem. Res.,* **2002**, *35*(9), 728-737.
[http://dx.doi.org/10.1021/ar010071i] [PMID: 12234202]

[62]   https://www.radiantinsights.com/research/glycerol-market

[63]   http://www.foodnavigator-usa.com/Financial-Industry/ADM-to-increase-US-sorbitol-production

[64]   Paster, M.; Pellegrino, J.L.; Carole, T.M. *Industrial Bioproducts: Today and Tomorrow*; Prepared by Energetics, Incorporated: Columbia, Maryland, **2003**.

[65]   Wood, A. Bioprocessing. *Chem. Week,* **2004**, *166*, 15-17.

[66]   http://www.tifac.org.in/offer/tsw/japenv.htm

[67]   Tate, B.E. *Itaconic Acid and Derivatives. Kirk-Othmer Encyclopedia of Chemical Technology,* 3rd ed; Grayson, M.; Eckroth, D., Eds.; John Wiley & Sons, **1981**, Vol. 13, .

[68]   Willke, T.; Vorlop, K-D. Biotechnological production of itaconic acid. *Appl. Microbiol. Biotechnol.,* **2001**, *56*(3-4), 289-295.
[http://dx.doi.org/10.1007/s002530100685] [PMID: 11548996]

[69]   Varma, I.K.; Albertsson, A.C. Enzyme catalyzed synthesis of polyesters. *Prog. Polym. Sci.,* **2005**, *30*, 949-981.
[http://dx.doi.org/10.1016/j.progpolymsci.2005.06.010]

[70]   Kelly, A.R.; Hayes, D.G. Lipase-Catalyzed Synthesis of Polyhydric Alcohol Poly(ricinoleic acid) Ester Star Polymers. *J. Appl. Polym. Sci.,* **2006**, *101*, 1646-1656.

[http://dx.doi.org/10.1002/app.22994]

[71]   Therisod, M.; Klibanov, A.M. Facile enzymatic preparation of monoacylated sugars in pyridine. *J. Am. Chem. Soc.,* **1986**, *108*, 5638-5640.
[http://dx.doi.org/10.1021/ja00278a053]

[72]   a) Patil, D.R.; Dordick, J.S.; Rethwisch, D.G. Chemoenzymatic synthesis of novel sucrose-containing polymers. *Macromolecules,* **1991**, *24*, 3462-3463.
[http://dx.doi.org/10.1021/ma00011a068] bKim, Dae-Yun; Dordick, J. S. Combinatorial array-based enzymatic polyester synthesis. *Biotechnol. Bioeng.,* **2001**, *76*, 200-206.c) Dordick, J.S. Enzymatic catalysis in monophasic organic solvents. *Enzyme Microb. Technol.,* **1989**, *11*, 194-211.
[http://dx.doi.org/10.1016/0141-0229(89)90094-X]

[73]   a) Ottosson, J.; Hult, K. Influence of acyl chain length on the enantioselectivity of Candida antarctica lipase B and its thermodynamic components in kinetic resolution of sec-alcohols. *J. Mol. Catal., B Enzym.,* **2001**, *11*, 1025-1028.
[http://dx.doi.org/10.1016/S1381-1177(00)00088-6] b) Malmstrom, E.; Johansson, M.; Hult, A. Hyperbranched Aliphatic Polyesters. *Macromolecules,* **1995**, *28*, 1698-1703.
[http://dx.doi.org/10.1021/ma00109a049]

[74]   Anderson, E.M.; Larsson, K.M.; Kirk, O. One Biocatalyst–Many Applications: The Use of Candida Antarctica B-Lipase in Organic Synthesis. *Biocatal. Biotransform.,* **1997**, *16*, 181-204.
[http://dx.doi.org/10.3109/10242429809003198]

[75]   Ballesteros, A.; Bornscheur, U.; Capewell, A.; Combes, D.; Condoret, J-S.; Koening, K.; Kolisis, F.N.; Marty, A.; Menge, U.K.; Scheper, T.; Stamatis, H.; Xenakis, A. Review Article Enzymes in Non-Conventional Phases. *Biocatalysis,* **1995**, *13*, 1-43.
[http://dx.doi.org/10.3109/10242429509040103]

[76]   Balcão, V.M.; Paiva, A.L.; Malcata, F.X. Bioreactors with immobilized lipases: State of the art. *Microb. Technol.,* **1996**, *18*, 192-416.
[http://dx.doi.org/10.1016/0141-0229(95)00125-5]

[77]   a) Kobayashi, S. Recent developments in lipase-catalyzed synthesis of polyesters. *Macromol. Rapid Commun.,* **2009**, *30*(4-5), 237-266.
[http://dx.doi.org/10.1002/marc.200800690] [PMID: 21706603] b) Kobayashi, S.; Makino, A. Enzymatic polymer synthesis: an opportunity for green polymer chemistry. *Chem. Rev.,* **2009**, *109*(11), 5288-5353.
[http://dx.doi.org/10.1021/cr900165z] [PMID: 19824647] c) Kobayashi, S.; Uyama, H.; Kimura, S. Enzymatic polymerization. *Chem. Rev.,* **2001**, *101*(12), 3793-3818.
[http://dx.doi.org/10.1021/cr990121l] [PMID: 11740921]

CHAPTER 7

Polymers Used as Catalyst

Arti Srivastava[*], **Bhaskar Sharma** and **Pratibha Mandal**

Department of Chemistry, Guru Ghasidas Vshwavidyalaya (A Central University), Koni, Bilaspur (Chhattsgarh) 495009, India

Abstract: In general, a catalyst is used to enhance the reaction and to complete the reaction quickly or accelerate the reaction involving reactants and catalysts. In the catalysis process, only the chemical structure of reactant changes with time, but the structure of the catalyst remains unaffected throughout the course of the reaction. The varieties of chemicals that can be used as a catalyst in numerous chemical reactions are metals, acids, bases, organic compounds, inorganic complexes, enzymes and polymers. Some specific polymers have the ability to catalyse reactions with the formation of carbon-carbon and carbon-non carbon linkages. Polyvinyl pyridine and sulfonated polystyrene are very useful and simple polymers that can act as catalysts. The catalytic activity of polymers is pronounced due to modification in polymer chains. Further, polymers may also be used as a support for another catalyst. Polymer catalysis can be illustrated with soluble linear polymers, ion exchange resins, polymer-supported phase transfer catalysts, palladium catalysts on polymer supports, *etc.* The brief review of each is explained by citing important examples along with their basic principles.

INTRODUCTION

The function of a catalyst is, to alter the rate of the reaction in order to form the product at a faster rate, and in the end, the catalyst will regenerate without any chemical changes. The catalysed reactions have a lower activation energy than uncatalysed reactions because catalyst abridged the potential energy barrier of each step of the main reaction. In this way, the catalyst provides the reduced height of the energy barrier of the split step. Eventually, the reactant can be converted into the final product with decreased Gibbs free energy without changing the overall standard Gibbs free energy of the reaction. The initial and final states of the reaction energetically remain the same. The catalytic processes are classified mainly into two types; (i) Heterogeneous and (ii) Homogeneous catalysis. In heterogeneous catalysis, the catalyst and reactants are in different phases. A typical example involves a solid catalyst with reactants as either in liquids or gases. In homogeneous catalysis, the catalyst remains in the same phase of reactants.

[*] **Corresponding author Arti Srivastava:** Department of Chemistry, Guru Ghasidas Vshwavidyalaya (A Central University), Koni, Bilaspur (Chhattsgarh) 495009, India; Tel: 07587448805; E-mail: artifeb@gmail.com

Goutam Kumar Patra & Santosh Singh Thakur (Eds.)

The activity of the substances that act as catalysts can be categorized as positive catalysts or simple catalysts, negative catalysts or inhibitors, self-catalyst or auto-catalyst and promoters. Positive catalysts are those catalysts, which enhance the rate of reaction while the negative catalysts are those catalysts, which reduce the rate of reaction. There are some reactions in which one of the reaction products acts as a catalyst and increases the rate of reaction, such type of catalysts are called self-catalyst or auto-catalyst. Another category of substances that are not themselves act as catalysts but increase the activity of a catalyst are called promoters.

Polymer catalysts have a basic skeleton of polymers possessing catalytically active moieties, which are usually attached to the side chain of synthetic polymers such as polystyrene. Some of the catalysts are incorporated into the main chain of the polymer. Cross-linked polymers are frequently used as polymer support material due to their insolubility. Insoluble polymer catalysts can be easily separated from the reaction mixture and reused.

Polymers in the field of catalysis occupied a large area of catalysis processes, which may be outlined through the important group of polymeric catalysts like synthetic hydrolases, immobilized enzymes, phase-transfer catalysts, nucleophilically active bases, polymers with conjugated π-systems, photosensitizers, and polymers as carriers and immobilized homogeneous catalysts. A lot of works have been reported *via* Palladium metal-based catalysts on polymer supports, polymer-bound Ru(III)–EDTA complex, which are now commercially available and used as catalysts in various organic reactions, including Suzuki-Miyaura, Mizoroki-Heck, Negishi, Stille, and Sonogashira coupling reactions. Chloromethylated styrene-divinylbenzene copolymer was chemically modified with an ethylenediaminetetraacetic acid ligand. The catalytically active polymer containing Ru(III) moieties were synthesized from this polymeric ligand and it was confirmed by elemental analyses, FTIR, UV, ESR and SEM [1].

This catalyst may be used up to two cycles for hydrogenation of 1-hexene to n-hexane and then recycled. Another such recent example is the microencapsulation of highly toxic Os species into the polymer. Continuous and quick works related to the development of smart catalysts have become an emerging field of controllable catalytic processes, *i.e.*, smart polymers for biomimetic catalysis and enzyme inhibition, responsive polymers as smart carriers in tunable catalytic processes, photocatalytic active polymers in organic synthesis, catalytic activity with polymer-based colloidal photonic crystals and porous polymer catalysts with hierarchical structures. M. Nasrollahzadeha *et al.* [2] published a detailed review on Palladium (Pd) complexes supported on the polymers, which are widely used in various organic synthesis like Sonogashira coupling reactions (SCR) of halobenzenes with alkynes due to their thermal stability, high catalytic activities and recyclability.

Catalysis by Soluble Linear Polymer

At the starting of the 1970s, Bayer's group has developed an alternative to insoluble polymer-bound reagents or catalysts that were prepared from the soluble polyethylene glycol or polystyrene derivatives as ligands, reagents or catalyst supports. In this respect, Bayer's group has first studied the synthesis of peptide and developed a methodology for peptide synthesis, which was termed a "liquid phase peptide synthesis" in which they used poly(ethylene glycol) as polymer support for the development of peptide chain. The development of a peptide chain using soluble polymer support has been carried out by following the basic principle of Merrifield, but the synthesis was performed in the homogeneous conditions. The intermediate products formed during peptide chain development have also been confirmed based on polymer-peptide's solubility. Bayer's group has also reported that both diphenyl phosphinated polystyrene and diphenyl phosphinated poly(ethylene glycol) could be used as recoverable, reusable hydroformylation catalysts. They have used properties of the polymer chain for isolation of products from the polymer-bound catalyst through solvent precipitation or membrane filtration techniques.

Catalysts are an important component of a reaction and generally, its cost is high, but still, it is necessary and difficult to separate from the product, which is another problem that raises the cost of the whole process high. A large number of different types of solid supports are available that are used to facilitate separation and reuse of catalyst [3 - 5]. For example, cross-linked polymers are used as flexible supports for organo-catalysts and ligands [6 - 9]. The high surface area and porous structure of the polymeric supports enhance the rate with quick diffusion of small molecules in and out [10, 11], while catalysts on the support diffuse very little. It was also reported by Kobayashi and others [12, 13] that catalysts attached to soluble polymers in the form of microcapsules will show high activity and can be easily tunable than the catalyst attached to the crosslinked resin. The model reaction that investigated is a 4-(N,N-dimethylamino)- pyridine (DMAP)-catalyzed acylation. DMAP is an ideal model because it has been studied as a small molecule [14] on linear polymers [15], inorganic supports [16], and insoluble polymeric supports. DMAP is also extremely sensitive to its electronic environment, as demonstrated by the extensive linker and backbone changes required to optimize its activity on a polymeric support. In contrast, it was shown that microencapsulated catalytic polymers can be optimized with minimal synthetic modification. The DMAP-modified linear polystyrene is formed by copolymerization of a DMAP-modified monomer and styrene.

Linear polymers with polar side groups have also been used as soluble catalysts. In this regard, Pinaud and co-workers reported syntheses of poly (vinylimidazolium salt) [17]. These imidazolium salts form corresponding *N*-heterocyclic carbenes (NHCs) on deprotonation that lead in turn to CO_2 adducts. These *N*-heterocyclic free carbenes are also used as polymer-bound catalyst and its concentration in reaction medium increases with increasing the temperature up to 80 °C. The polymer-bound catalyst can be recycled by adding the reaction mixture to excess ether.

Soluble Polymer-supported Organocatalyst

Recently, enormous modifications have been studied with the development of organocatalysts, which provide the new route for chemical transformations. Some drawbacks came into notice that many of the chemical transformations accompanied by low turnover rates and frequencies. Therefore, recyclisation and reuse of the organocatalyst and resultant immobilization of catalyst onto soluble polymer support are the main keys for the successful use of polymer-supported organocatalyst. The soluble polymer-supported organocatalysts can be designed easily and characterized by solution-state spectroscopy after their successful synthesis. In catalysis by soluble polymer-supported organocatalyst, the catalyst and product both are in the same phase, therefore sometimes, it was separated by permselective membranes, but it was not successful and still not widely used for recovery/reuse of soluble polymer bound catalysts. Several other methods have been used to separate the product from soluble polymer bound catalysts like soluble polymer support can be separated from the soluble product by precipitating the soluble polymer bound catalysts. Both may also be separated by heating and cooling. In spite of several above options, many more have opted for the separation and reuse of catalysts such as selective polymer precipitation, filtration, centrifugation, by perturbation of solution of the soluble polymer-bound catalyst and product so as to turn the solution into two liquid phases, gravity separation, *etc.*

The polar and/or non-polar soluble polymers (Fig. **1**) have been used as support with organocatalysts [18].

 a. Polyisobutylene (PIB), (b) Poly(ethylene glycol) (PEG), (c) Polyethylene, (d) polystyrene, (e) Alkylated polystyrene, (f) Dendrimer

(a) Polyisobutylene (PIB), (b) Poly(ethylene glycol) (PEG), (c) Polyethylene, (d) polystyrene

(e) Alkylated polystyrene   (f) Dendrimer

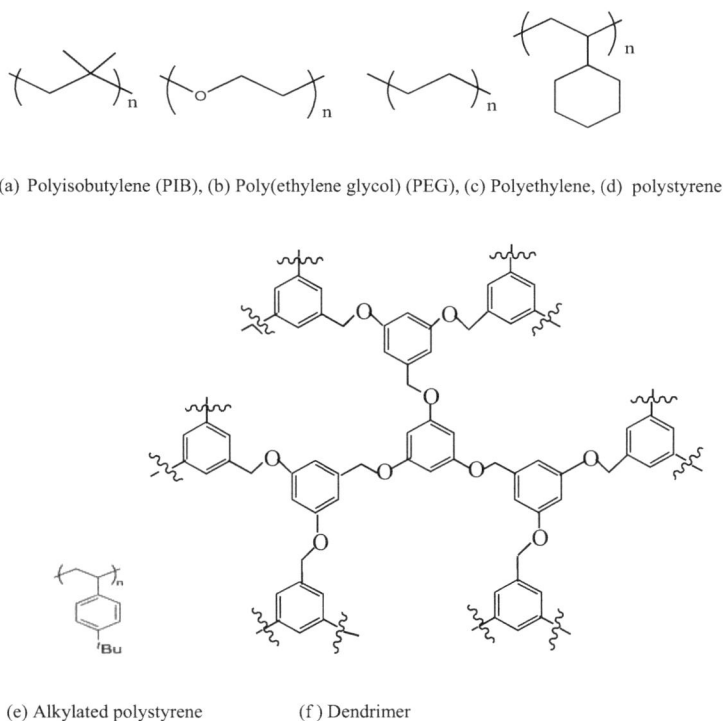

Fig. (1). Chemical structure of soluble polymer supports.

Polyisobutylene (Fig. **1a**) is a nonpolar support and it is used with transition-metal catalysts and also with organocatalysts. The allylation of benzaldehydes has been carried out by using PIB-supported pyridine N-oxide catalyst [19]. The gravity separation technique has been used to separate the catalyst/product mixture. The PIB-bound catalyst/product mixture has been dissolved in a biphasic mixture of hexane/90% then PIB-bound catalyst came in hexane; hence, in this way catalyst can be isolated after hexane removal and reused. This catalyst was reused for five cycles with good conversions without significant loss of activity. This result suggests that hydrocarbon polymers like PIB should be generally useful as supports for recoverable organocatalysts.

Bergbreiter group in 2011 has reported syntheses of the PIB-bound alkyldiphenyl phosphine and triaryl phosphine [20]. These polymer-supported phosphines organocatalysts have been used in alkyne addition and allylic substitution reactions. These catalysts are efficient and recyclable.

The alkyne addition reaction can also be done with isobutyldiphenyl phoshine.

This catalyst is low molecular weight and electronically similar to the PIB-bound alkyl diphenylphosphine. It has been observed that phosphine catalysts are very efficient catalysts proved with the formation of high yield of product mixture and these catalysts are recyclable and reusable and give good yield up to 5 cycles beyond that it will give a low yield of the product mixture [21].

Jørgensen [22, 23] and Hayashi [24] in 2005 independently reported that diarylprolinol silyl ethers are efficient and recyclable organocatalysts, which are used in varieties of the chemical reaction.

In 2010, Mager and Zeitler reported a PEG-supported catalyst. This catalyst has been prepared through copper-catalyzed Fokin–Huisgen reaction of a PEG-azide and a propargyl diarylprolinol silyl ether [25].

The polymer-bound L-proline catalyst has been used in Michael addition reaction of cyclohexanone to nitrostyrene using MeOH as the solvent and its chemical structure has been represented by Fig. (2). The reported yield of the product is 92%, with 46% ee. The same reaction has also been carried out by (2S, 4S)-4-acetoxyproline catalyst that is a low-molecular-weight catalyst. The reported yield of the product has been found to be 89%, with 40% ee. The PEG-bound catalyst has been separated by precipitation and filtration with the addition of ether in reaction mixture. In this case, the PEG-bound catalyst was recycled and reused three times. However, significant decreases in yield from 94% in the first run to 24% in the fourth run was observed.

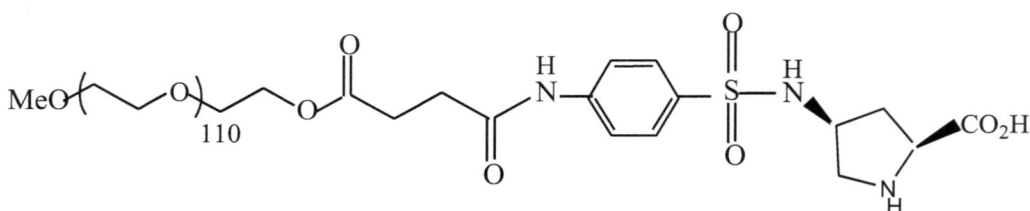

Fig. (2). Chemical Structure of PEG-polymer-bound L-proline catalyst.

Dendrimers are the functional soluble polymers, which are used as supports for catalysts and reagents [26]. The functionalized **dendrimers** compromise the potential of both homogeneous and heterogeneous **catalysis**. The dendritic **catalysts** can show the activity and selectivity of a conventional homogeneous **catalyst**, and they can be separated easily from the reaction medium. This is illustrated by the work of Bellis and Kokotos, who prepared L-proline-modified poly (propyleneimine) dendrimers with five generations [27]. These L-prolin--modified dendrimers, along with L-proline hydrochloride, have been used as catalyst precursors in the aldol reaction of acetone and 4-nitrobenzaldehyde.

Triethylamine was used as a base to generate proline moieties *in situ*. Among the different generations of dendrimer-supported proline catalysts, the second-generation proline-modified dendrimer showed the best results. The aldol product was isolated in 70% yield with 53% ee. Reactions using dendrimers with higher generations with the same proline loading provided 60–68% yields of the product but with lower enantioselectivity (21–41% ee).

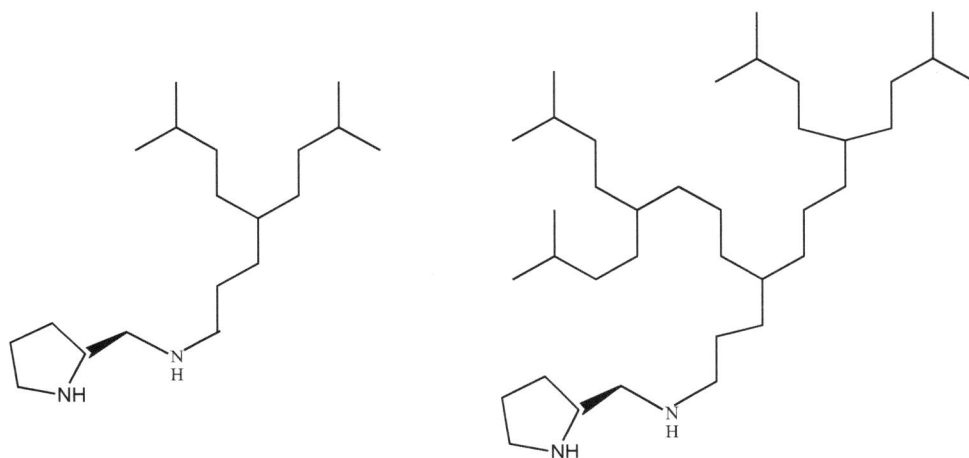

Fig. (3). Chemical Structure of Dendrimer-supported diamino organocatalysts.

The dendrimer-supported diamino organocatalysts are phase selectively soluble in the heptane phase in a biphasic heptane/MeOH mixture (Fig. **3**). Recycling of dendrimers was possible in a nitro-Michael reaction in which the crude product mixture was dissolved in a mixture of heptane and MeOH and the catalyst was selectively soluble in the heptane phase and the product mixture was selectively soluble in the MeOH phase. It may be reused four times with the same yields and stereoselectivity as in the first cycle. These Catalysts were also used in aldol reactions. In these cases, similar yields and stereoselectivity were also observed in the product mixtures throughout five cycles.

Catalysis by Ion Exchange resins

An ion-exchange resin or ion-exchange polymer is a resin or polymer originated from an organic substrate, which is an insoluble matrix (or support structure). It is a white or yellowish microbead and the size of these microbeads ranges from **0.25–0.5 mm**. An ion exchange resin is an insoluble polymer matrix having labile ions, which have nature to exchange with ions in the surrounding medium and the physical and chemical state of resin will remain as such after the exchange of ions. Based on the nature of ions, ion exchange resin is of two types, namely cation and anion exchangers. Those polymeric resins which are anionic in nature can be used as cations exchange resins. The anion exchange resins are polymeric cations which have a labile, exchangeable anion.

In the functionality point of view, the resins are known as cation exchange resins if they contain sulfonic, carboxylic, and phenolic groups and they are known as anion exchange resins if the functional group is a quaternary ammonium or ammonium group.

In 1905, the German chemist Gans used sodium aluminosilicate materials (*zeolites*) to soften water. About 600 different zeolites are known and frequently used in the purification of polluted water. It is an inorganic cation exchanger which has good cation exchange selectivity and excellent compatibility with the environment. It is frequently used as selective adsorbents, molecular sieves, and especially as catalysts due to its high selectivity, non-toxicity, high resistance to heat and ion radiation. Zeolite minerals are used in the treatment of polluted water that comes out from nuclear wastewater, municipal and industrial wastewaters, acid mine drainage waters and other construction materials.

The first synthetic ion exchange resins were developed by Adams and Holmes in 1935, which is based on a phenol-formaldehyde structure. After that, the cross-linked polystyrene matrix has been developed, which covers 99% use of industrial resin, and recently, polyacrylic resins have emerged in the market, which has broadened the scope and versatility of the synthetic ion exchange resins, and covers the remaining 10% of the commercial market. The synthetic polystyrene and polyacrylic acid have been used in the form of a spherical bead that provides the more contact surface to the solution in comparison to the irregular shaped phenol-formaldehyde resin.

The cross-linked polymer of polystyrene and divinylbenzene was developed in 1944. The crosslinked copolymer of styrene and dibenzene contains about 6%–8% divinylbenzene, and the rest is covered by vinyl styrene. It is observed that crosslinked copolymeric resin has many times more ion exchange capacity than other earlier developed resins when taken with an equal volume of the resins. Later on, its analog, *i.e.*, anion exchange resin, was developed in 1948. The complete demineralization of water has been completed by using both resins. Many modifications of the original polystyrene divinylbenzene have been made to accomplish a wide variety of separations.

The first gel-type polystyrene resin has been developed in 1947 in which exchange takes place by diffusion of the ions through the resin.

In 1956, Mikes and co-workers discovered the concept of pores or holes into the resin structure due to which a new type of polymeric materials developed that is called as macroporous resin. New dimensions in the technique of ion exchange resin broaden with the development of macroporous polymeric resins. These resins encompass a continuous polymer matrix intermingled with a continuous pore matrix.

Olefin Oligomerization

Amberlyst, Purolite, and Nafion are the commercial name of ion exchange resins. Lot of chemically different types of Amberlyst, Purolite, and Nafion have been used as catalyst in oligomerization of many olefines like propene, butene, isobutene and isoamylene *etc.* oligomerization of C_4 alkenes have been carried out with the use of Amberlyst®-type of resins [28] and oligomerization of higher olefins (C_{10-32}) with Nafion® resin. These oligomers can be used as diesel fuels and as lubricant after hydrogenation. The solid acid-catalyzed dimerization of methylstyrene (AMS) has been reported by Chakrabarti and Sharma using a number of ion-exchange resins like clays and Nafion®-NR50 resin as catalysts in non-polar solvent cumene and also in polar solvent *p*-cresol at 50°C [29, 30]. The liquid-phase synthesis of 2-ethoxy-2-methylpropane (ETBE) and 2-ethoxy-2-methylbutane (TAEE) has been studied over fifteen commercial acidic ion-exchange resins [31]. The catalytic activity of commercial acidic ion-exchange resins for the synthesis of 2-ethoxy-2-methylpropane (ETBE) and 2-ethoxy-2-methylbutane (TAEE) have been measured as intrinsic initial etherification rates and it has been found to decrease in the order: Amberlyst™ 35 > Amberlyst™ 48 > Purolite® CT-275 > Amberlyst™ 15 > Purolite® CT-175 > Amberlyst™ 40 > Amberlyst™ 36 > Amberlyst™ 16 > Purolite® CT-482 > Amberlyst™ 39 > Amberlyst™ DT > Amberlyst™ 45 > Purolite® CT-124 > Purolite® MN-500 > Amberlyst™ 46.

This catalytic activity order has been given on the basis of their morphological properties in dry and swollen states. Another parameter has been used to determine the order of their catalytic activity *i.e.*, the ratio of acid capacity to the specific volume of the swollen polymer. It has been found that the catalytic activity is directly related to the ratio that means higher the ratio, higher the activity.

Etherification via Alcohol Addition to Olefins

Ethers are known as oxygenated gasoline additives, which are used to improve the quality of petrol by reducing the emission of fuel from motor vehicles. The most commonly used ethers are methyl *t*-butyl ether (MTBE), ethyl *t*-butyl ether

(ETBE) and *t*-amyl methyl ether (TAME). They are synthesized by the addition of alcohol to olefins using an acid catalyst [32 - 35]. MTBE is used as a replacement for lead in the US gasoline pool for a very long time. The MTBE is synthesized from 2-methylpropene (isobutene) and methanol in excess using an anionic ion-exchange sulfonic acid resin catalyst, *e.g.*, Amberlyst®-15, Dowex® M32 [36] between 340-360 K and at 8 atm pressure. The Ethyl t-butyl ether (ETBE) is an alternative to MTBE as an oxygenate to enhance the octane rating of petrol and this covers 90% of the world's annual production of ca 3 million tonnes. Ethyl tert-butyl ether is manufactured industrially by the acidic etherification of isobutylene with ethanol at a temperature of 30 °C – 110 °C and a pressure of 0,8–1,3 MPa catalyzed by an acidic ion-exchange resin catalyst. The t-Amyl methyl ether (TAME) is another alternative to MTBE. It is prepared from 2-methyl but-1-ene and 2-methyl but-2-ene during the various refinery processes to make petrol, including catalytic cracking. A plug flow reactor is used to mix the alkenes with methanol and pass over an ion-exchange resin such as a co-polymer of a sulfonated phenylethene and divinylbenzene. The sulfonated groups, $-SO_3H$, provide the acidic groups which catalyse the reaction. It is generally supposed that the activity of the resin catalyst depends on its sulfonic acid capacity. It has been observed that the higher the acid capacity, the higher the catalytic activity [37]. Firstly, the catalytic activity of macroporous type resin catalyst depends on its acid groups and next on the surface area and pore diameter. The macroporous resin is preferred due to the high accessibility of its acid groups [38]. Other ethers that are also used as fuel additives such as diisopropyl ether [39], *t*-butylpropyl ether [40], and so on can be synthesized with cation-exchange resin catalyst [41, 42].

Carbon-Carbon Coupling Reactions Catalyzed by Nanoparticles Supported on Commercial Ion-Exchange Resins

The carbon-carbon coupling reaction may be catalyzed by Pd, Ag, Au, metals nanoparticles, which support ion exchange resins and nanoparticles, which have been synthesized by the reduction method. The metal loading is dependent on the ion-exchange capacity of the resin and depends upon the size and charge of the metal ion. Charged functional groups on the resin and pores help in stabilizing the nanoparticles. Some recent works based on these catalysts are as follows [43]:

- Reetz and coworkers have reported the Suzuki reaction with palladium nanoparticles stabilized by R_4N^+X and, in another case, stabilized by ionic polymers with a quaternary ammonium group. Liang *et al.* reported a metal-organic chemical vapor deposition (MOCVD) approach to load Amberlite IRA-900(OH), a basic ion-exchanger, with a volatile palladium precursor

Pd(C_3H_5)(C_5H_5). The different sizes of Palladium nanoparticles have been reported with different methods like 20 nm PdNPS have been reported from N_2 adsorption and 2.6 nm with the chemical method. It was observed that vapor deposition methods give better dispersion of nanoparticles in the resin matrix. An equal volume of Ethanol/water was used as a solvent. The catalyst was used five times, with a slight decrease in the activity. They also mentioned that when the chloride form of the ion-exchanger was replaced with an OH^- form, it lost its catalytic activity [44].

• In another report, core-shell of Pd/Co bimetallic nanoparticles supported on a sulfate ion-exchange resin were present near the surface of the resin. They were found to be active for the cross-coupling of 4-bromo acetophenone with phenylboronic acid in DMF/water using K_2CO_3 as a base [45]. Macronet MN100, a hyper cross-linked PS-DVB resin functionalized with tertiary amine groups, was used to synthesize palladium nanoparticles by the sorption reduction method. $PdCl_2$ was used as a precursor and reduced with sodium formate and sodium hydroxide in a solution. The catalyst was used to study the Suzuki reaction. The catalyst was shown to have good efficiency for aryl bromides and chlorides using Cs_2CO_3 as a base [46].

• Suzuki and Heck are the most common reactions explored by polymer-stabilized nanoparticles, Sengupta *et al.* explored the feasibility of a Sonogashira reaction using bimetallic Pd/Cu nanoparticles. Cationic macroporous Amberlite resin with formate as a counter anion was used to support bimetallic Pd/Cu nanoparticles. The Co-impregnation method was used to synthesize well-dispersed nanoparticles with an average size of 4.9 nm. The synthesized catalyst exhibited high catalytic activity for Sonogashira cross-coupling reaction between aryl iodide and alkynes in CH_3CN at 80°C. No homocoupled products were detected in the reaction mixture [47].

Polymer Resins as Nanoreactors for the Synthesis of Nanoparticles and Their Catalytic Application in C-C Coupling

We shall consider polymeric organic catalysts containing metals in this section. These will be divided into two groups: in the first group, the catalytic centers contain free metal; in the second, the metal is bound to the polymer in the form of an ion, a complex, or a chelate.

Catalysts Containing Free Metal

At first sight, the group of polymers containing free metal appears to correspond to the heterogeneous catalysts on inorganic carriers-a group used widely in industry-but with the difference that the carrier is a polymer. Here already, the analogy breaks down because organic polymers can assume varied chemical or physical forms and be suited to a specific purpose. For example, metal particles with a mean diameter of about 15 nm can be produced in a polymer gel:

a metal salt solution is equilibrated with a neutral, slightly cross-linked gel of poly(2-hydroxyethyl methacrylate) and then reduced with sodium tetrahydridoborate. Since the swelling capacity of the gel varies according to the solvent, the selectivity can be controlled by diffusion for these nonpyrophoric, readily separable, and re-usable catalysts. This has been demonstrated, among other cases, for the hydrogenation of cinnamaldehyde, mainly to hydrocinnamaldehyde, in the presence of a Pd-containing gel; the selectivity is higher than with a conventional Pd/C catalyst. On the other hand, PtO_2, Ni, or Rh particles can also be microencapsulated by cross-linking interfacial polycondensation; if ferromagnetic material is added to the batch, then after hydrogenation the capsules can be easily separated from the batch by a magnetic field. As a third possibility, the catalyst particles may be dispersed in solid polymers that can then be processed into some form, *e.g.,* film, which is advantageous for a technical process.

Catalysts Containing the Metal in the form of a Compound, Ion, or Complex

Polymers in general, but particularly ion exchangers and polymers with ligands that form complexes or chelates, are also well suited as carriers for catalytically active metal complexes. The advantages of these catalysts are again that they can be easily separated and re-used while an initially completely soluble, low-molecular (homogeneous) catalyst with increasing concentration reaches saturation and thus an activity independent of the amount present, for a high-molecular catalyst-a strongly acid cation exchanger charged with $Pd(NH_3)$ a linear relationship is found between its activity and its amount over a wide region. Saturation phenomena are thus avoided, and in such cases, an increased rate of conversion is attained. Homogeneous catalysts are clearly defined substances that can be prepared for a high degree of reproducibility. Since the steric and electronic environment of the catalytic centers in homogeneous catalysts can be varied within wide limits, the selectivity and the reaction rate can be controlled. There is no way to apply to heterogeneous catalysts, which are, therefore, difficult to develop further. In addition, in the latter, only a small proportion of the active centers are accessible and hence effective, whereas, in the homogeneous phase in principle, all active molecules take part in the course of the reaction. Furthermore, homogeneously catalyzed reactions can often be carried out under milder conditions and have better mass and temperature characteristics. In conjunction with these clear advantages of homogeneous catalysts, however, some problems that arise from the industrial application must also be mentioned. In the first place, some of the noble-metal complexes have a corrosive action on the reactors, leading to deposition of metal on the reactor walls; this results in a loss of the catalyst and in the formation of new active centers of undesirable properties. Apart from the difficulty in realizing a continuous mode of the reaction, however, the main disadvantage is the difficulty of the catalyst's recovery. According to the nature of the catalyst and the product, this necessitates a repetition of expensive separations. Moreover, in many cases, the active catalyst cannot be recovered as such, but only a portion of the noble metal.

Recent advances in Polymer-supported Pd (PSP) Complexes for Sonogashira Coupling Reactions (SCR)

The **Sonogashira reaction** was given by Sonogashira *et al.* in 1975. This is a cross-coupling reaction and basically well known for the formation of carbon-carbon bonds in an organic reaction. It employs a palladium catalyst as well as copper co-catalyst to form a carbon-carbon bond between a terminal alkyne and an aryl or vinyl halide. This reaction is applied in the polymer and pharmaceutical industries, material science and in the preparation of heterocyclic compounds, biologically active substances and natural products.

Palladium complexes supported on polymer matrix are used as a complex heterogeneous catalyst in the various named reaction. These catalysts are widely used due to their excellent thermal and chemical stabilities and product selectivity. The immobilization of the palladium complexes in the polymer matrix enhances the stability of this heterogeneous complex catalyst at a higher temperature in comparison to analogous homogeneous catalysts. Such types of catalysts cannot spoil with an increase in pressure and temperature conditions. These catalysts can be easily synthesized using various ligands and used several times. This catalyst can also be used after long time storage with the same catalytic activity efficiency as used in the first cycle. Mahmoud Nasrollahzadeh *et al.* [48] reviewed the synthesis of heterogeneous catalyst complexes based on different organic compounds with Polymer-support and Pd metal and its application in **Sonogashira coupling reaction.** Brief about some catalysts are as follows:

- Macrocyclic Schiff base-based PSP complex was introduced as a novel, robust and thermally stable catalyst [49] and exhibited high catalytic performance in the SCR with only a 0.5 mol % Pd loading. This catalyst system can be reused five times with high catalytic activity.
- Another PCP Schiff base complex was prepared by reacting polystyrene amine with benzaldehyde after 72 h reflux in dry toluene solvent. The synthesized polymer Schiff base ligand with palladium acetate in AcOH at 80 °C for 10 h catalyze the SCR under ligand- and Cu-free conditions using Et$_3$N and TBAB in

DMF/H$_2$O mixture. This catalyst could be reused in sequential six cycles.

- The fabrication of polymeric NHC grafted silica-supported Pd NPs (Si-PNH--Pd) have been reported by Tamami *et al.* [50] This unique Si-PNHCPd catalytic system has been used to catalyzed SCR with significant activity. The catalyst was consecutively reused 12 times. The thermal stability and high recyclability are its great features.

- The synthesis of superparamagnetic NPs (γ-Fe$_2$O$_3$/polymer) supported dendritic catalyst based on phosphine Pd (II) complex [51] and the assembly of a catalyst formed by the reaction of carboxylic acid groups of the γ-Fe2O3/polymer with the amino [52]. The complex catalyzed the copper-free SCR in the presence of a surfactant where the colloidal dispersion of surfactant particles played a significant part in the cross-coupling yield.

- Thiosemicarbazone-based PSP complex [PS-ppdot-Pd(II)] was introduced by Bakherad *et al.* [53]. This thiosemicarbazone-based PSP complex [PS-ppdo--Pd(II)] was synthesized by reacting 1-phenyl-1,2-propanedione-2-oxime thiosemi-carbazone (PPDOT) and chloromethylated polystyrene at 100 °C for 20 h. Then, PPDOT functionalized polymer refluxed with [PdCl$_2$(PhCN)$_2$] using ethanol. The synthesized PS-ppdot-Pd(II) catalyst applied to catalyzed SCR using Et$_3$N or pyridine as the base at ambient temperature with good product yield. The catalyst could be reused for four successive cycles with good efficiency. Further Thiosemicarbazone-based PSP complex **[PS-PPdot-Pd(0)]** was introduced by Bakherad *et al.* [54]. This catalyst can be prepared by heating the [PS-ppdot-Pd(II)] complex with N$_2$H$_4$.H$_2$O at 50 °C for 5 h in ethanol. This copper- and solvent-free SCR using [PS-PPdot-Pd(0)] and Et$_3$N at ambient temperature was explored with high product yields. The complex was reused for several cycles. In the same way, phenyl dithiocarbazate-based PSP complex was also synthesized by reacting phenylhydrazine with carbon disulfide and chloromethylated polystyrene using KOH at ambient temperature in 3 h.

- S. M. Islam *et al.* [55] synthesized a novel catalyst (PS-Pd (II)-furfural complex based on furfural with macroporous amino polystyrene and subsequent complexation with Pd(OAc)$_2$. The coupling reactions of different aryl halides with organoboronic acid, alkene, amine and alkyne have been carried out by using furfural-functionalized polymer grafted with Pd(II)) complex catalyst. It was found that PS-Pd (II)-furfural complex to be an inexpensive, stable and recyclable catalyst for the SCR. It could be used with the same efficiency up to five times.

- The SMI-PdCl$_2$ complex is a novel, reusable and efficient catalyst [56], which was synthesized by supporting the PdCl$_2$ on modified poly(styrene-co-maleic anhydride) (SMA) with 2-aminothiazolen. The ensuing catalyst SMI-PdCl$_2$ catalyzed SCR of aryl halides and various terminal acetylenes. Another efficient catalyst is Fe$_3$O$_4$@SiO$_2$-polymer-imid-Pd magnetic complex and applied in the

SCR [57] and could be reused for six successive cycles. Merrifield resin (PS) reinforced air-stable Pd NHC complex was explored in the presence of a spacer (Pd-NHC@SP-PS) and absence of a spacer (Pd-NHC@PS) [58]. These complexes were synthesized *via* immobilization of an ionic liquid on the surface of the polymer and the reaction of the ensuing amended polymer with Pd salt. The functionalized PS-SP resin was prepared by reacting 6-chlorohexanol with commercially available PS using NaH in THF for 48 h.

Polymeric Phase-transfer Catalysts and Nucleophilically Active Polymers

Phase transfer catalysis came in light from 1965. This technique becomes a very useful method of synthetic organic chemistry because, with the introduction to phase transfer catalysis, the problem of synthetic organic synthesis was solved in which reacting reagents are not able to contact each other for useful reactions. This difficulty of bringing together a water-soluble nucleophilic reagent and an organic water-insoluble electrophilic reagent was improved by the addition of a solvent, which is both water-like and organic-like, for example, ethanol. Phase-transfer catalysts (PTC) are those catalysts, which help in transferring reactant from one phase into another phase in the reaction medium through a catalytic phase transfer agent. Phase-transfer catalysis comes under the category of heterogeneous catalysis. PTC is widely used in the synthesis of various organic chemicals in both liquid-liquid and solid-liquid systems. Polymer-supported phase transfer catalysts are insoluble polymers in which the active part that acts as catalysts is covalently bounded, functional groups. The active functional groups may be quaternary ammonium or phosphonium ions, crown ethers, cryptands, grafted poly (ethylene glycols), or analogues of dipolar aprotic solvents. The polymer most often used is polystyrene, but other synthetic polymers, silica gel, and alumina have also been used.

The phase transfer catalysis makes the reaction faster with high conversion of reactant to products with reduced side products in spite of these PTC desires low costs and biocompatible solvents that make the Phase-transfer catalysts as a green catalyst [59, 60].

There are many types of phase transfer catalysts, such as quaternary ammonium and phosphonium salts, crown ethers and cryptands. The quaternary ammonium salts are widely used due to its low cost. There are two basic types of PTC, including crown ethers and cryptands.

The former include compounds such as 18-crown-6 (1) dibenzo-18-crown-6 (2) dicyclohexano-18-crown-6 (3) crypands such as the [2.2.2.] compound (4) and terminally capped and uncapped oligooxyethylenes (Fig. **4**).

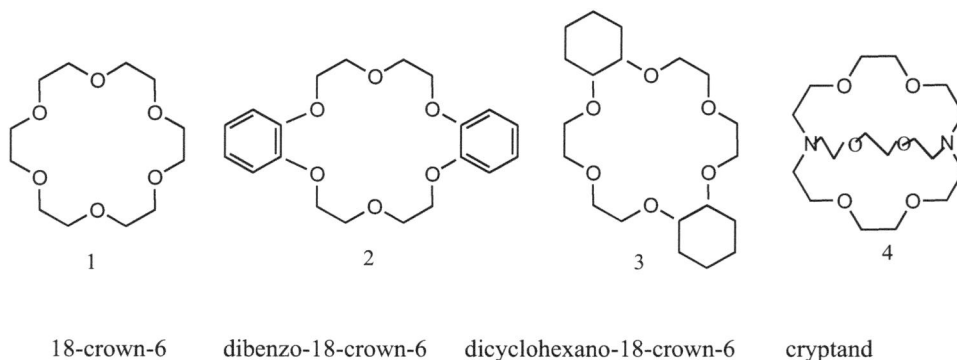

1	2	3	4
18-crown-6	dibenzo-18-crown-6	dicyclohexano-18-crown-6	cryptand

Fig. (4). Structure of different Crown Ethers.

Phase Transfer Catalyst has great importance to catalyze the various reactions in which inorganic and organic anions, carbenes and other species react with organic compounds. In recent years several different asymmetric transformations catalyzed by chiral onium salts and crown ethers have been used in the synthesis of valuable organic compounds. Phase-transfer catalyst has been used in lots of important organic synthesis because of their distinctive properties such as operational simplicity, mild reaction conditions, appropriateness for large-scale synthesis, and benevolent environment of the reaction system.

The linear and crosslinked functional polymers have catalytic applications in phase transfer reactions. Polymers containing either quaternary ammonium group or oligoether moieties (glymes and crown ethers) in the side chain, experience a leading chirality. The study of chemical kinetics and stereochemical control for the reduction of prochiral alkyl ketones, alkylation of phenylacetonitrile and oxidation of racemic alcohols have been done for the analysis of polymer-supported phase transfer catalyst activity. The catalysts have been immobilized on a polymer to overcome the problem related to the separation of the catalyst from the final product in the purification process of two soluble phases. The polymer-supported catalysts and reagents can be easily separated from the reaction mixture by simple filtration technique and can be reused the catalyst.

The poly (glycidyl methacrylate) (PGMA) has two reactive hydroxyl groups, and this Macroporous PGMA resin is used as a support for phase-transfer catalysts, due to the probability of functionalization by direct attachment or the generation of a variety of functional groups. It provides an environment that is more polar than that afforded by the more usual styrene-divinylbenzene resins [61]. The Macroporous PGMA supported phase-transfer catalysts used in organic synthesis of a series of tetraalkyl ammonium and tetraaryl phosphonium salts. Some exclusive properties, which arises due to the attachment of PTC's with polymers, are described as:

Polymer Modification and Functionalization

The main area of PTC applications to polymers involves the chemical modification of functional polymers using two-and three-phase systems. The former may involve liquid-liquid or liquid-solid reactions, while the latter generally involves two liquid phases interacting with an insoluble polymer.

In the early stage of PTC development, Roovers investigated the use of PTC in polymer modifications in 1976 [62]. The substrate polymer was soluble, chloromethylated polystyrene and reactions have been carried in 50:50 benzene acetonitrile at 75 °C with either dicyclohexano-18-crown-6 (3) or dibenzo-1--crown-6 (2) as the PTC. The former was about twice as efficient as the latter under these conditions. The different potassium carboxylates nucleophiles and acid anions like acetate (100%); benzoate (100%); trans-cinnimate (87%); and l-phenylpropionate (100%) have been used in the formation of corresponding bound esters (Fig. **5**). The latter served as a model system for the reaction of mono-carboxylate-terminated polystyrene (Mn=3000 daltons). The complete conversion to graft or comb-polymer has been done with this procedure.

The modification of Polyepichlorohydrin and copolymer has been done with ethylene oxide and by assimilation of carbaeole (Fig. **6**) using 5 mole % of tetrabutyl-ammonium PTC at 60°C. The 31% highest conversions have been obtained on both polymers using DMT as the solvent. The conversion has been increased up to 56%, with increasing the catalyst concentration on 100 mole %. It has also been observed that there is no significant increase in conversion to change of other reaction conditions and concentrations.

Three-phase catalysis has been applied to the generation of polymer-bound sulfonium ylids, which were then used to form epoxides (Fig. **7**) [63]. The dimethyl or diethyl sulfonium intermediate was converted to the ylids and used in epoxide synthesis using DMSO as a solvent.

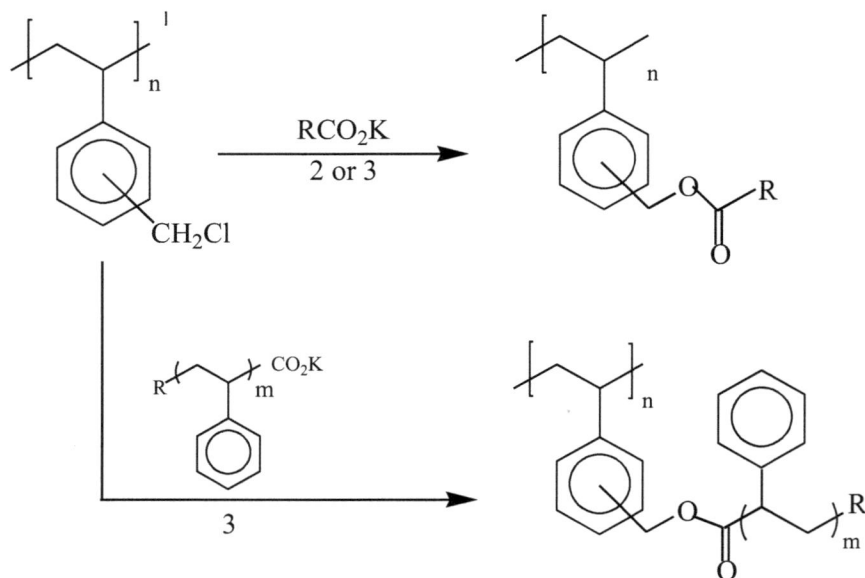

Fig. (5). Chloromethylated polystyrene and potassium carboxylates formed the corresponding bound esters using PTC.

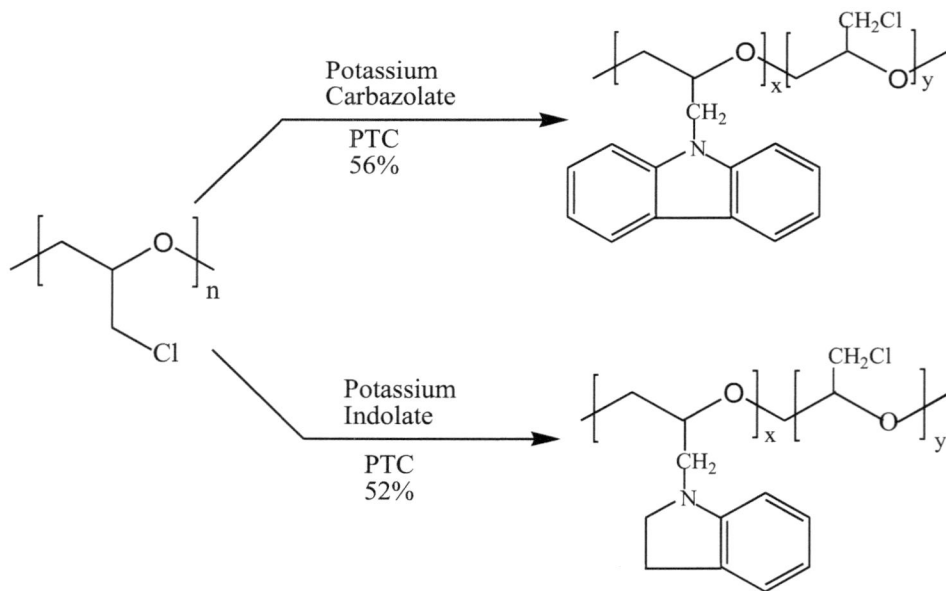

Fig. (6). Modification of Polyepichlorohydrin using PTC.

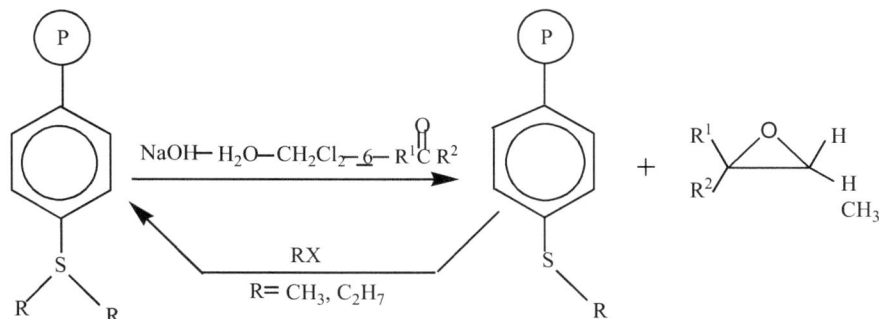

Fig. (7). Generation of polymer-bound sulfonium ylids (Three-phase catalysis).

Further, the conversion of chloromethylated polystyrene to dinitriles and diamine has been later completed by an extensive examination of similar compounds by another group [64]. Frechet and his colleagues treated insoluble chloromethylated polystyrene with the anion of malononitrile under three-phase conditions to give the polymer-bound dinitrile in essentially complete conversion (Fig. **8**).

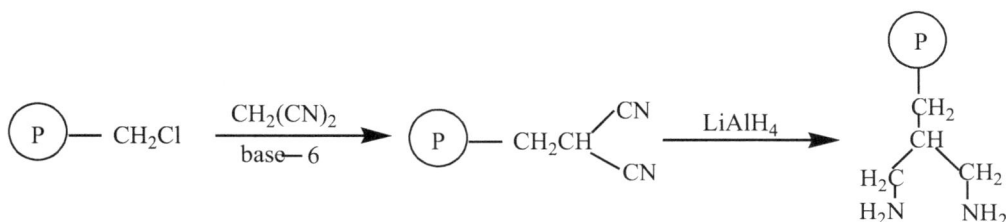

Fig. (8). Conversion of chloromethylated poly-styrene to dinitriles and diamine.

Ricard and co-workers reported a slightly different procedure for incorporation of the 1, 3-diaminopropane group (Fig. **9**). Acylation of the polystyrene with chloroacetyl chloride was followed by a three-phase reaction with malononitrile directly or after the initial reduction of the ketone group to an alcohol. The former gave a much higher overall yield of the polymer-bound diaminoalcohol [65].

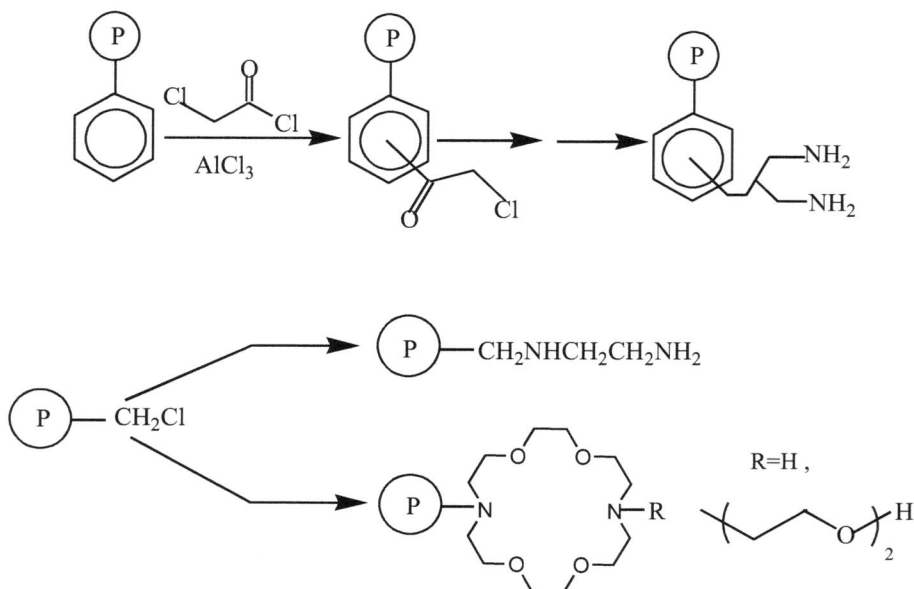

Fig. (9). Polymer-bound diaminoalcohol formation using PTC with high yield.

CONSENT FOR PUBLICATION

Not applicable.

CONFLICT OF INTEREST

The author declares that there is no conflict of interest in this chapter.

ACKNOWLEDGEMENTS

Declared none.

REFERENCES

[1]   Dalal, M.K.; Ram, R.N. Catalytic activity of polymer-bound Ru(III)-EDTA complex. *Bull. Mater. Sci.,* **2001,** *24*(2), 237-241.
[http://dx.doi.org/10.1007/BF02710108]

[2]   Nasrollahzadeha, M.; Motahharifara, N.; Ghorbannezhada, F.; Bidgolia, N.S.S.; Baranb, T.; Varmac, R.S. Recent advances in polymer supported palladium complexes as (nano)catalysts for Sonogashira coupling reaction. *Molecular Catalysis,* **2020,** *480*: 110645.
[http://dx.doi.org/10.1016/j.mcat.2019.110645]

[3]   McNamara, C.A.; Dixon, M.J.; Bradley, M. Recoverable catalysts and reagents using recyclable polystyrene-based supports. *Chem. Rev.,* **2002,** *102*(10), 3275-3300.
[http://dx.doi.org/10.1021/cr0103571] [PMID: 12371885]

[4]   Ley, S. V.; Baxendale, I. R.; Bream, R. N.; Jackson, P. S.; Leach, A. G.; Longbottom, D. A.; Nesi, M.; Scott, J. S.; Storer, R. I.; Taylor, S. J. Multi-step organic synthesis using solid-supported reagents and scavengers: a new paradigm in chemical library generation. *J. Chem. Soc., Perkin Trans.,* **2000**, 3815.

[5]   Barrett, A.G.M.; Hopkins, B.T.; Köbberling, J. ROMPgel reagents in parallel synthesis. *Chem. Rev.,* **2002**, *102*(10), 3301-3324.
[http://dx.doi.org/10.1021/cr0103423] [PMID: 12371886]

[6]   Selkala, S.A.; Tois, J.; Pihko, P.M.; Koskinen, A.M.P. Asymmetric Organocatalytic Diels–Alder Reactions on Solid Support. *Adv. Synth. Catal.,* **2002**, *344*, 941.
[http://dx.doi.org/10.1002/1615-4169(200210)344:9<941::AID-ADSC941>3.0.CO;2-M]

[7]   Benaglia, M.; Puglisi, A.; Cozzi, F. Polymer-supported organic catalysts. *Chem. Rev.,* **2003**, *103*(9), 3401-3429.
[http://dx.doi.org/10.1021/cr010440o] [PMID: 12964876]

[8]   Parrish, C.A.; Buchwald, S.L. Use of polymer-supported dialkylphosphinobiphenyl ligands for palladium-catalyzed amination and Suzuki reactions. *J. Org. Chem.,* **2001**, *66*(11), 3820-3827.
[http://dx.doi.org/10.1021/jo010025w] [PMID: 11375003]

[9]   Leadbeater, N.E.; Marco, M. Preparation of polymer-supported ligands and metal complexes for use in catalysis. *Chem. Rev.,* **2002**, *102*(10), 3217-3273.
[http://dx.doi.org/10.1021/cr010361c] [PMID: 12371884]

[10]   Gambs, C.; Dickerson, T.J.; Mahajan, S.; Pasternack, L.B.; Janda, K.D. High-Resolution Diffusion-Ordered Spectroscopy To Probe the Microenvironment of JandaJel and Merrifield Resins. *Org Chem.,* **2003**, *68*, 3673.
[http://dx.doi.org/10.1021/jo034152z]

[11]   Yamane, Y.; Kobayashi, M.; Kuroki, S.; Ando, I. Diffusional Behavior of Solvents and Amino Acids in Network Polystyrene Gels As Studied by 1H Pulsed-Field-Gradient Spin–Echo NMR Method. *Macromolecules,* **2001**, *34*, 5961.
[http://dx.doi.org/10.1021/ma010320i]

[12]   Kobayashi, S.; Akiyama, R. Renaissance of immobilized catalysts. New types of polymer-supported catalysts, 'microencapsulated catalysts', which enable environmentally benign and powerful high-throughput organic synthesis. *Chem. Commun. (Camb.),* **2003**, (4), 449-460.
[http://dx.doi.org/10.1039/b207445a] [PMID: 12638949]

[13]   He, HS.; Yan, J.J.; Shen, R.; Zhou, S.; Toy, P.H. Non-Cross-Linked Polystyrene-Supported Triphenylphosphine-Microencapsulated Palladium: An Efficient and Recyclable Catalyst for Suzuki-Miyaura Reactions. *Synlett,* **2006**, *4*, 563-566.

[14]   Ragnarsson, U.; Grehn, L. Novel Amine Chemistry Based on DMAP-Catalyzed Acylation. *Acc. Chem. Res.,* **1998**, *31*, 494.
[http://dx.doi.org/10.1021/ar980001k]

[15]   Bergbreiter, D.E.; Osburn, P.L.; Li, C. Soluble polymer-supported catalysts containing azo dyes. *Org. Lett.,* **2002**, *4*(5), 737-740.
[http://dx.doi.org/10.1021/ol017198s] [PMID: 11869115]

[16]   Chen, H.T.; Huh, S.; Wiench, J.W.; Pruski, M.; Lin, V.S. Dialkylaminopyridine-functionalized mesoporous silica nanosphere as an efficient and highly stable heterogeneous nucleophilic catalyst. *J. Am. Chem. Soc.,* **2005**, *127*(38), 13305-13311.
[http://dx.doi.org/10.1021/ja0524898] [PMID: 16173762]

[17]   Pinaud, J.; Vignolle, J.; Gnanou, Y.; Taton, D. Poly(N-heterocyclic-carbene)s and their CO2 Adducts as Recyclable Polymer-Supported Organocatalysts for Benzoin Condensation and Transesterification Reactions. *Macromolecules,* **2011**, *44*, 1900.
[http://dx.doi.org/10.1021/ma1024285]

[18]   Yang, Y.C.; Bergbreiter, D.E. Soluble polymer-supported organocatalysts. *Pure Appl. Chem.,* **2013**,

85(3), 493.
[http://dx.doi.org/10.1351/PAC-CON-12-05-03]

[19]   Bergbreiter, D.E.; Yang, Y-C.; Hobbs, C.E. Polyisobutylene-supported phosphines as recyclable and regenerable catalysts and reagents. *J. Org. Chem.*, **2011**, *76*(16), 6912-6917.
[http://dx.doi.org/10.1021/jo201097x] [PMID: 21714575]

[20]   Bergbreiter, D.E.; Chandran, R. Polyethylene-bound rhodium(I) hydrogenation catalysts. *J. Am. Chem. Soc.*, **1987**, *109*, 174.
[http://dx.doi.org/10.1021/ja00235a027]

[21]   Marigo, M.; Wabnitz, T.C.; Fielenbach, D.; Jørgensen, K.A. Enantioselective Organocatalyzed α Sulfenylation of Aldehydes. *Angew. Chem. Int. Ed.*, **2005**, *44*, 794.
[http://dx.doi.org/10.1002/anie.200462101]

[22]   Franzén, J.; Marigo, M.; Fielenbach, D.; Wabnitz, T.C.; Kjaersgaard, A.; Jørgensen, K.A. A general organocatalyst for direct alpha-functionalization of aldehydes: stereoselective C-C, C-N, C-F, C-Br, and C-S bond-forming reactions. Scope and mechanistic insights. *J. Am. Chem. Soc.*, **2005**, *127*(51), 18296-18304.
[http://dx.doi.org/10.1021/ja056120u] [PMID: 16366584]

[23]   Hayashi, Y.; Gotoh, H.; Hayashi, T.; Shoji, M. Diphenylprolinol Silyl Ethers as Efficient Organocatalysts for the Asymmetric Michael Reaction of Aldehydes and Nitroalkenes. *Angew. Chem. Int. Ed.*, **2005**, *44*, 4212.
[http://dx.doi.org/10.1002/anie.200500599]

[24]   Mager, I.; Zeitler, K. Efficient, enantioselective iminium catalysis with an immobilized, recyclable diarylprolinol silyl ether catalyst. *Org. Lett.*, **2010**, *12*(7), 1480-1483.
[http://dx.doi.org/10.1021/ol100166z] [PMID: 20201561]

[25]   Ouali, A.; Caminade, A-M. *Dendrimers: Towards Catalytic, Material, and Biomedical Uses*; John Wiley, **2011**, p. 163.

[26]   O'Connor, C.T.; Kojima, M.; Schumann, W.K. The oligomerization of C4 alkenes of cationic exchange resins. *J. Catal.*, **1985**, *16*, 193.

[27]   Sun, Q.; Farneth, W.E.; Harmer, M.A. Dimerization of α-Methylstyrene (AMS) Catalyzed by Sulfonic Acid Resins: A Quantitative Kinetic Study. *J. Catal.*, **1996**, *164*, 62.
[http://dx.doi.org/10.1006/jcat.1996.0363]

[28]   Chakrabarti, A.; Sharma, M.M. Some novel aspects of dimerization of .alpha.-methylstyrene with acidic ion-exchange resins, clays and other acidic materials as catalysts. *Ind. Eng. Chem. Res.*, **1989**, *28*, 1757.
[http://dx.doi.org/10.1021/ie00096a004]

[29]   Yadav, G.D.; Krishnan, M.S. An Ecofriendly Catalytic Route for the Preparation of Perfumery Grade Methyl Anthranilate from Anthranilic Acid and Methanol. *Org. Process Res. Dev.*, **1998**, *2*, 86.
[http://dx.doi.org/10.1021/op970047d]

[30]   Abro, S.; Pouilloux, Y.; Barrault, J. Selective synthesis of monoglycerides from glycerol and oleic acid in the presence of solid catalysts. *Stud. Surf. Sci. Catal.*, **1997**, *108*, 536.
[http://dx.doi.org/10.1016/S0167-2991(97)80948-2]

[31]   Soto, R.; Fite, C.; Ramirez, E.; Iborra, M.; Tejero, J. Catalytic activity dependence on morphological properties of acidic ion-exchange resins for the simultaneous ETBE and TAEE liquid-phase synthesis. *React. Chem. Eng.*, **2018**, *3*, 195.
[http://dx.doi.org/10.1039/C7RE00177K]

[32]   Bart, H.J.; Kaltenbrunner, W.; Landschuetzer, H. Kinetics of esterification of acetic acid with propyl alcohol by heterogeneous catalysis. *Int. J. Chem. Kinet.*, **1996**, *2*, 649.
[http://dx.doi.org/10.1002/(SICI)1097-4601(1996)28:9<649::AID-KIN2>3.0.CO;2-V]

[33]   Saha, B.; Streat, M. Transesterification of cyclohexyl acrylate with n-butanol and 2-ethylhexanol:

acid-treated clay, ion exchange resins and tetrabutyl titanate as catalysts. *React. Funct. Polym.,* **1999**, *40*, 13.
[http://dx.doi.org/10.1016/S1381-5148(98)00004-2]

[34]   Panneman, H.J.; Beenackers, A.A.C.M. Synthesis of Methyl tert-Butyl Ether Catalyzed by Acidic Ion-Exchange Resins. Influence of the Proton Activity. *Ind. Eng. Chem. Res.,* **1995**, *34*, 4318.
        [http://dx.doi.org/10.1021/ie00039a023]

[35]   Parra, D.; Izquierdo, J.F.; Cunill, F.; Tejero, J.; Fite, C.; Iborra, M.; Vila, M. Catalytic Activity and Deactivation of Acidic Ion-Exchange Resins in Methyl tert-Butyl Ether Liquid-Phase Synthesis. *Ind. Eng. Chem. Res.,* **1998**, *37*, 3575.
        [http://dx.doi.org/10.1021/ie980007d]

[36]   Buttersack, C. Accessibility and catalytic activity of sulfonic acid ion-exchange resins in different solvents. *React. Poly,* **1989**, *10*, 143.
        [http://dx.doi.org/10.1016/0923-1137(89)90022-5]

[37]   Heese, F.P.; Dry, M.E.; Moller, K.P. Single stage synthesis of diisopropyl ether – an alternative octane enhancer for lead-free petrol. *Catal. Today,* **1999**, *49*, 327.
        [http://dx.doi.org/10.1016/S0920-5861(98)00440-4]

[38]   Calderon, A.; Tejero, J.; Izquierdo, J.F.; Iborra, M.; Cunill, F. Equilibrium Constants for the Liquid-Phase Synthesis of Isopropyl tert-Butyl Ether from 2-Propanol and Isobutene. *Ind. Eng. Chem. Res.,* **1997**, *36*, 896.
        [http://dx.doi.org/10.1021/ie960492h]

[39]   Zhang, T.J.; Jensen, K.; Kitchaiya, P.; Phillips, C.; Datta, R. Liquid-Phase Synthesis of Ethanol-Derived Mixed Tertiary Alkyl Ethyl Ethers in an Isothermal Integral Packed-Bed Reactor. *Ind. Eng. Chem. Res.,* **1997**, *36*, 4586.
        [http://dx.doi.org/10.1021/ie970099r]

[40]   Linnekoski, J.A.; Krause, A.O.I.; Struckmann, K. Etherification and hydration of isoamylenes with ion exchange resin. *Appl. Catal. A,* **1998**, *170*, 117.
        [http://dx.doi.org/10.1016/S0926-860X(98)00040-4]

[41]   Samahe Sajade. *Encapsulated catalysts*; Published by Academic Press, Elsevier, **2017**, pp. 1-537.
        [http://dx.doi.org/10.1016/B978-0-12-803836-9.00001-8]

[42]   Zhang, M.; Jiang, M.; Liang, C. Palladium supported on an ion exchange resin for the Suzuki-Miyaura reaction. *Chin. J. Catal.,* **2013**, *34*, 2161.
        [http://dx.doi.org/10.1016/S1872-2067(12)60698-6]

[43]   Alonso, A.; Shafir, A. Recyclable polymer-stabilized nanocatalysts with enhanced accessibility for reactants. j. macans, A. Valeribera, M. Munoz, D. Muraviev, catal. *Today,* **2012**, *193*, 200.
        [http://dx.doi.org/10.1016/j.cattod.2012.02.003]

[44]   lyubimov, S.E.; Vasilev, A.A.; Korlyukov, M.M.; Ilyin, S.A.; Pisarev, V.V.; Matveev, Palladium containing hypercrosslinked polystyrene as an easy to prepare catalyst for Suzuki reaction in water and organic solvents. *react. Funct. Polym,* **2009**, *69*, 755.

[45]   Sengupta, D.; Saha, J.; De, G.; Basu, B. Pd/ Cu bimetallic nanoparticles embedded in macroporous ion- exchange resins: an excellent heterogeneous catalyst for the Sonogashira reaction. *J. Mater. Chem,* **2014**, *A 2*, 3986.

[46]   Zhang, W.; Peng, L.; Deng, C.; Zhang, Y.; Qian, H. . *Mol. Catal.,* **2018**, *445*, 170.
        [http://dx.doi.org/10.1016/j.mcat.2017.11.036]

[47]   Chen, X.; Wang, W.; Zhu, H.; Yang, W.; Ding, Y. A simple method for preparing imidazolium-based polymer as solid catalyst for Suzuki-Miyaura reaction. *Mol. Catal.,* **2018**, *456*, 49.
        [http://dx.doi.org/10.1016/j.mcat.2018.07.007]

[48]   Nasrollahzadeha, M.; Motahharifara, N.; Ghorbannezhada, F.; Bidgolia, N.S.S.; Baranb, T.; Varmac, R.S. Recent advances in polymer supported palladium complexes as (nano)catalysts for Sonogashira

coupling reaction. *Molecular Catalysis,* **2020**, *480*110645
[http://dx.doi.org/10.1016/j.mcat.2019.110645]

[49] He, Y.; Cai, C. Heterogeneous copper-free Sonogashira coupling reaction catalyzed by a reusable palladium Schiff base complex in water. *J. Organomet. Chem.,* **2011**, *696*(13), 2689.
[http://dx.doi.org/10.1016/j.jorganchem.2011.04.012]

[50] Tamami, B.; Farjadian, F.; Ghasemi, S.; Allahyari, H. Synthesis and applications of polymeric N-heterocyclic carbene palladium complex-grafted silica as a novel recyclable nano-catalyst for Heck and Sonogashira coupling" reactionsHYPERLINK \l "fn1. *New J. Chem.,* **2013**, *37*, 2011.
[http://dx.doi.org/10.1039/c3nj41137k]

[51] Rosario-Amorin, D.; Gaboyard, M.; Clerac, R.; Nlate, S.; Heuze, K. Enhanced catalyst recovery in an aqueous copper-free Sonogashira cross-coupling reaction. *Dalton Trans.,* **2010**, 40.

[52] Islam, M.; Mondal, P.; Roy, A.S.; Tuhina, K.; Mondal, S.; Hossain, D. Polystyrene-Anchored Palladium(II) Schiff Base Complex: A Reusable Catalyst for Phosphine-Free and Copper-Free Sonogashira Cross-Coupling Reaction in Aqueous Medium. *Synth. Commun.,* **2011**, *41*, 2583.
[http://dx.doi.org/10.1080/00397911.2010.515331]

[53] Bakherad, M.; Keivanloo, A.; Bahramian, B.; Jajarmi, S. Copper- and solvent-free Sonogashira coupling reactions of aryl halides with terminal alkynes catalyzed by 1-phenyl-1,2-propanedio-e-2-oxime thiosemi-carbazone-functionalized polystyrene resin supported Pd(II) complex under aerobic conditions. *Appl. Catal. A Gen.,* **2010**, *390*(1-2), 135.
[http://dx.doi.org/10.1016/j.apcata.2010.10.003]

[54] Bakherad, M.; Keivanloo, A.; Bahramian, B.; Jajarmi, S. Synthesis of Ynones via Recyclable Polystyrene-Supported Palladium(0) Complex Catalyzed Acylation of Terminal Alkynes with Acyl Chlorides under Copper- and Solvent-Free. *Synlett,* **2011**, *3*, 311.
[http://dx.doi.org/10.1055/s-0030-1259322]

[55] Islam, S.M.; Salam, N.; Mondal, P.; Roy, A.S. Highly efficient recyclable polymer anchored palladium catalyst for CC and CN coupling reactions. *J. Mol. Catal. Chem.,* **2013**, *366*, 321.
[http://dx.doi.org/10.1016/j.molcata.2012.10.011]

[56] Heravi, M.M.; Hashemi, E.; Shirazi Beheshtiha, Y.; Ahmadi, S.; Hosseinnejad, T. PdCl$_2$ on modified poly(styrene-co-maleic anhydride): A highly active and recyclable catalyst for the Suzuki–Miyaura and Sonogashira reactions. *J. Mol. Catal. Chem.,* **2014**, *394*, 74.
[http://dx.doi.org/10.1016/j.molcata.2014.07.001]

[57] Esmaeilpour, M.; Javidi, J.; Mokhtari Abarghoui, M.; Nowroozi Dodeji, F. Fe$_3$O$_4$@SiO$_2$-polyme--imid-Pd magnetic porous nanosphere as magnetically separable catalyst for Mizoroki–Heck and Suzuki–Miyaura coupling reactions. *J. Iran. Chem. Soc.,* **2014**, *11*, 499.
[http://dx.doi.org/10.1007/s13738-014-0443-5]

[58] Jadhav, S.N.; Kumbhar, A.S.; Mali, S.S.; Kook Hong, C.; Salunkhe, R.S. A Merrifield resin supported Pd–NHC complex with a spacer(Pd–NHC@SP–PS) for the Sonogashira coupling reaction under copper- and solvent-free conditions. *New J. Chem.,* **2015**, *39*, 2333.
[http://dx.doi.org/10.1039/C4NJ02025A]

[59] Metzger, J.O. Solvent-Free Organic Syntheses. *Angew. Chem. Int. Ed.,* **1998**, *37*, 2975.
[http://dx.doi.org/10.1002/(SICI)1521-3773(19981116)37:21<2975::AID-ANIE2975>3.0.CO;2-A]

[60] Makosza, M. Phase-transfer catalysis. A general green methodology in organic synthesis. *Pure Appl. Chem.,* **2000**, *72*, 1399.
[http://dx.doi.org/10.1351/pac200072071399]

[61] Kenawy, E.L-R. Polymer-supported phase-transfer catalysts: synthesis and high catalytic activity of ammonium and phosphonium salts bound to linear and crosslinked poly(glycidyl methacrylate). *Des. Monomers Polym.,* **1998**, *1*(2), 155.
[http://dx.doi.org/10.1163/156855598X00297]

[62]   Roovers, J.E.L. Crown ether catalysed modification of partly chloromethylated polystyrene. *Polymer (Guildf.)*, **1976**, *17*, 1107.
[http://dx.doi.org/10.1016/0032-3861(76)90016-1]

[63]   Guyen, T.D.N.; Deffieux, A.; Boileau, S. Phase-transfer catalysis in the chemical modification of polymers. *Polymer (Guildf.)*, **1978**, *19*, 423.
[http://dx.doi.org/10.1016/0032-3861(78)90249-5]

[64]   Frechet, JMJ.; Smet, MD.; Farrall, MJ. Chemical modification of crosslinked resins by phase transfer catalysis: preparation of polymer-bound dinitriles and diamines. *Tetrahedron Letters.*, **1979**, *20*(2), 137-138.

[65]   Ricard, M.; Villemfn, D.; Ricard, A. Preparation of polymer-bound 1,2 - and 1,3 – diamines. *Tetrahedron Letters,* **1980**, *21*(1), 47-50.

SUBJECT INDEX

A

Acetals 1, 11, 13, 157
  benzofused 11, 13
  bicyclic 1
  cyclic ketene 157
Acetates 54, 154, 181, 185
  palladium 181
Acetic acid 70
Acid-base interactions 77, 78
  non-covalent 78
Acid capacity 177, 178
  sulfonic 178
Acid catalysts 5, 150, 153, 178
  imidodiphosphoric 5
Acids 2, 3, 4, 10, 12, 18, 23, 33, 36, 71, 77,
    78, 108, 127, 140, 148, 150, 154, 157,
    169, 176, 179, 182
  achiral phosphoric 3
  benzoic 78
  carboxylic 23, 33, 36, 157
  chiral phosphoric 2, 4, 140
  derived dicarboxylic 159
  fumaric 148
  organic sulfonated 77
  organoboronic 182
  peroxyformic 150
  phenylboronic 179
  phosphotungstic 71
  polyacrylic 176
  solid 77
  sulfamic 108
  trichloroacetic 127
  trifluoroacetic 71, 77
  unsaturated fatty 154
Activation 16, 25, 30, 42, 48, 53, 67, 93, 123,
    125, 137, 140, 142
  cooperative 42, 53
  dual 142
  mode of 142
  preferential 16
Activation energies 54, 169
  lower 169
  minimum 54

Active sites 67
Active substances 56, 181
  valuable 56
Addition 12, 41, 97, 172, 173
  alkyne 173
  asymmetric catalytic phenyl 72
  catalyzed conjugated 12
  oxa-Michael 97
  surfactant 41
Agricultural economies 161
Alcohol addition 177
Alcohols 2, 82, 83, 96, 115, 150, 156, 178,
    187
  sugar 156
Alcoholysis 18
Aldehydes 72, 73, 74, 75, 77, 98, 99, 100,
    108, 131, 132, 133, 134, 135
  aliphatic 75
  aromatic 74, 108
  enantioselective arylation of 72
  linear 77, 133
Alkylation reactions 114
Arylation reaction 72
  first asymmetric 72
Asymmetric arylation 72, 73
  of aldehydes 73
Asymmetric 25, 31, 33, 35, 36, 50, 52, 55, 67,
    80, 82, 83, 84, 138
  carbene transfer reactions 83, 84
  catalysts 55
  cyanation 80
  cyanosilylation 52
  heterogeneous catalysis 67, 80
  kinetic resolution 25, 50
  phase-transfer catalysis 138
  ring-opening (ARO) 25, 31, 33, 35, 36
  Stetter reaction 82
Asymmetric Michael 77, 78, 79
  addition reaction 78, 79
  reactions 77
Asymmetric synthesis 2, 9, 16, 17, 18, 56,
    124, 125, 126, 127, 142
  catalytic 9
  large-scale 17

petrol-derived 159
petroleum-based 159
toxic 149, 161
Chemoselective 10, 105, 128
  catalyst control 128
  switch 10
Chhattisgarh council of science and
  technology (CCOST) 56
Chiral 1, 2, 3, 4, 56, 131, 138, 140
  onium salts 184
  organofluorine compounds 131
  organometallic catalysts 56
  phase-transfer catalysts 138
  phosphoric acid (CPA) 2, 3, 4, 140
  quaternary ammonium salts 138
  spiro compounds 1
Chiral catalysts 21, 29, 36, 42, 55, 56, 67, 69,
  78, 80
  developed 80
  efficient 55
  efficient novel 56
  pre-prepared 69
  stable 78
Chiral fluorinating 135, 136
  agent 136
  reagents 135
Cinchona alkaloids 71, 127, 129, 135, 136,
  137, 139, 141
  combination of 135, 136
Cinchona-based alkaloid catalysts 127
Classical Michael addition reaction 96
Click reactionss 109
Co-impregnation method 179
Commercial biobased polyol-prepolymer 151
Commercial ion-exchange resins 178
Complexes 16, 36, 83
  activated 36
  bimetallic chiral salen Co 16
  polystyrene-supported Pybox-Ru 83
Complicated synthesis method 29
Condensation 74, 85, 105, 107, 134, 155, 160,
  175
  asymmetric aldol 75
  base-catalyzed 105
  copolymerization 160
  cycloaddition reactions 107
  intramolecular aldol 85
  polymerizations 155
  tetrazolecatalyzed aldol 74

Condensation reactions 93, 98, 100, 102, 103,
  106, 118, 155, 156, 161
  demonstrated solid-state mechanochemical
    cascade 106
  lipase-catalyzed 155, 156
  selective 156, 161
Corynebacterium glutamicum 160
Cutinase-catalyzed biotransformation 157
Cycles 133, 140
  chiral organocatalytic 140
  organocatalytic 133
Cyclic 2, 48, 157
  diesters 157
  frameworks 2, 48
Cyclisation reaction methods 14
Cycloaddition reactions 1, 14, 102, 112, 113
  demonstrated Diels-Alder 102
Cyclocondensation reactions 105

D

Dehydration 46, 74, 154, 159
  oxidative 159
Dehydrogenations 66
Density functional theory (DFT) 37, 48, 53,
  54
Deposition 52, 178, 181
  diffusion-limited atomic layer 52
  metal-organic chemical vapor 178
Deprotonation 172
Derivatives 73, 113
  pyrrolidine 73
  reacting amines 113
DIANANE-salen complexes 41
Diastereoisomers 10, 17, 74, 75
  purified 17
  syn chiral 75
Diastereoselective acetalization 12
Dicarboxylic acid 160
Diels-Alder cycloaddition reactions 102
Drawbacks of exiting polyol technology 154
Dynamic kinetic resolution (DKR) 27, 46, 49

E

Effect 35, 48
  cooperative catalysis 48
  strong synergistic 35
Efficient catalyst immobilization 72

www.ingramcontent.com/pod-product-compliance
Lightning Source LLC
Chambersburg PA
CBHW050845220326
41598CB00006B/434